The Author

Andrew Caine

He graduated with honours in Marine Biology from University of North Wales, Bangor.

The author of two successful books:

Marine Biology for the Non-Biologist
Marine Ecology for the Non-Ecologist

He has been extensively published in various marine-related magazines over a period of ten years. During this time, he worked in various areas within the marine industry, concentrating on tropical marine biology.

He also worked in remote Indonesia on coral reef regeneration projects, teaching the local population how to cultivate corals in their natural environment and to help repair damaged coral reefs.

He has now settled down to concentrate on his family and is teaching and inspiring the next generation of scientists.

You can always reach out by email to:

andrewmcaine@gmail.com

Dedications

This book is dedicated to two fantastic people,

Tracey, and Matthew.

Both have put up with my trials and tribulations throughout our lives and they have both provided unquestionable support. Without this, my books would not have been possible, and I would be a different person than I am today.

For once in my life, I am truly speechless!

Two other amazing people I would like to thank are Mark and Corinna for providing the peaceful and tranquil environment I needed, where I could be left alone.

All rights reserved. No part of this publication may be reproduced, stored in a retrieval system, or transmitted in any form or by any means, electronic, mechanical, photocopying, recording, or otherwise without the prior permission of the copyright owner.

©2021 by Andrew Caine

Author
Andrew Caine

Editor
Eve Cunard

Co-ordinator and sponsor
Matthew Caine

Incredible Oceans!

Amazing facts and explanations from the wonderful worlds of

Marine biology
Marine ecology
and
Oceanography

*To my favourite ADDICT
Andy*

INTRODUCTION

Before we start, I would like to explain what this book is not. It is not an endless stream of sentences and short paragraphs providing only facts. That sort of publication is indeed fantastic and fun; however, I would like to expand by not only providing amazing facts about the oceans and the life it contains, but also by including explanations to cement a degree of understanding.

Yes, we have chapters to compartmentalise everything- it happens in all walks of life. Nonetheless, as with anything biological, there are always exceptions to the rule. Our joint exploration into the wonderful world of the oceans is no different. Thrown in, just for the fun of it, within the chapters, are amazing facts that I cannot put into their own categories or chapters. I could not leave them out due to their fantastic nature. Enjoy!

So this is a book full of incredible facts and explanations about the oceans and the life it contains. Yet, if we look at the oceans and our planet as a whole, we could only have one starting point: the birth of the universe.

The universe was formed (or we believe it was formed) around 13.7 billion years ago—give or take a couple of years. This had very little to do with the earth as we know it; for the first 9 billion years we quite simply did not exist.

However, something crucial to the existence of future life on earth was a very special molecule that did occur within the universe: water, found mainly in its solid form, ice.

As we know, the universe is full of stars. They are factories producing atoms and elements, so the atoms that make up your body were indeed produced in a star before our solar system existed.

When a star dies, depending on many factors, it undergoes a process that throws out all of these atoms and elements into the galaxy they were once part of. The universe is full of dead stars and what they have left behind. They are like clouds in the sky, but instead of being made up of water, they are made up of this cosmic dust.

Luckily for us, billions of years ago, one such cloud started to collapse in on itself through the force of gravity. As all the particles in this cloud rubbed together, they got hotter through friction. Then 4.5 billion years ago, pressure and temperature got so extreme that the atoms started to join together, nuclear fusion occurred, and our star exploded into existence.

The sun took 99.9% of the cosmic cloud for itself, while the remaining 0.1% formed the planets, moons and all the other stuff in our solar system.

The earth formed, and during the first 800 million years something very special happened: comets and asteroids kept hitting the earth. Each time one crashed into our young piece of rock it brought ice, and slowly, over time, the universe topped up our planet with water. Our oceans were born.

So, when did life start?

Answer 1) 3.8 billion years ago (or longer) around hydrothermal vents on the ocean floor.

Answer 2) 3.7 billion years ago in the earth's surface oceans.

We have no evidence for answer one—it's just a thought, but we do have evidence for answer two.

3.7 billion years ago, the earth's atmosphere would not support life as we know it today because it was full of carbon dioxide and methane. Oxygen was hardly there at all.

The first life on earth was bacteria that we know nothing about. Yet, they left behind a form of carbon only produced by living organisms. We have evidence of that carbon in rocks that are 3.7 billion years old.

Our oldest fossils are the houses made by bacteria called stromatolites, dating from 3.5 billion years ago. We were on our way.

Only 2.4 billion years ago, a cell evolved that produced its own food and its waste product was oxygen. However, it took another 1.6 billion years for these cells to transform our atmosphere into one that held enough oxygen for more complex life to exist.

The oldest animal fossil known is a 700 million-year old sponge. This indicates that the origin of animal life was in the oceans. It existed here for another 350 million years, before the first amphibians crawled onto land; a land that was already full of trees and plants.

Fast forward to the present - plenty of evolution has developed the amazing life we see in the oceans today. So much so, that it is estimated that 99.9% of all species that ever lived on earth are now extinct. Don't be alarmed, this is natural; when one species evolves, another species dies out. It's purely natural.

There are 226 000 or so marine species alive today, but there could quite easily be over a million - we don't know.

So, there is only one thing to do, let's go and find out about them.

Contents

1. THE OCEAN

IT'S ONE BODY OF WATER AROUND
THE GLOBE AND IT'S AMAZING   Pages 11 - 30

2. THE PLANKTON

MOSTLY INVISIBLE TO HUMANS,
LIFE-GIVING, YET OFTEN DEADLY
                                    Pages 31 - 46

3. LOCOMOTION AND MIGRATION

FROM SLOW TO SUPER-SPEEDY,
IT'S ALL A MATTER OF LIFE AND DEATH

                                    Pages 47 - 64

4. FEEDING

FINE DINING TO FLUID ONLY—
NO TABLE MANNERS HERE   Pages 65 - 63

5. REPRODUCTION AND LIFE SPANS

PASSING ON DNA OR JUST MAKING COPIES, WITH MIND-
BLOWINGLY DIVERSE STRATEGIES.

LIFESPANS FROM HOURS TO HUNDREDS OF YEARS
                                    Pages 84 -101

6. HOUSING

THE SUPPLY AND DEMAND OF OCEAN REAL ESTATE
Pages 102 - 117

7. RELATIONSHIPS BETWEEN SPECIES

WHO HELPS WHO AND WHY?   Pages 118 - 125

8. WHATS GOING TO KILL YOU?

CREATURES TO AVOID;
EATING ONE IS DEADLY TOO   Pages 126 - 143

9. THREATS TO THE OCEANS

IS THE SOLUTION TO POLLUTION DILUTION?
Pages 144 – 163

10. THE MARINE LIFE SERIES
Pages 164 - 169

THE OCEAN

EARLY OCEANS

The last megaflood occurred between 5.5 and 4.5 million years ago when the Atlantic Ocean simply poured into the Mediterranean Sea.

Let us go a little further back in time, say 6 million years, when the earth's crust had moved to form a barrier at the mouth of the Mediterranean Sea. This cut off the sea from the Atlantic Ocean. Over thousands of years, evaporation was at a far greater rate than rain and the Mediterranean simply dried out. It became a huge valley of desiccated salt; in some places the valley bottom was 3 km deep with 1.5 km-high sheer cliffs.

Then, with the movement of the earth's crust, the barrier opened and the Atlantic Ocean poured in. Such was the force of this water that it carved a 200 km-long channel hundreds of metres deep into the valley floor, cascading over steep cliffs and forming a waterfall over 1.5 km high. The Mediterranean Sea filled up at a rate of over 10 m in depth per day. In under two years, the Mediterranean was at its current level. It was the most violent flood in earth's history.

13 000 years ago, sea level was 125 m (410 ft) lower than it is today and the last great extinction of life on earth occurred with a meteorite striking Canada.

We know how old the earth is, and to understand our oceans we only have to rewind 13 000 years. At this time, the earth was between 5 and 10° C cooler — yes, we were in the last ice age. The earth was about to undergo its last major shift, both in habitat construction and in the life that existed here.

Great ice sheets and glaciers covered the northern parts of Europe, Canada, Greenland, and North America. Basically, this whole northern area was frozen. Although the southern hemisphere also showed great ice formations, the majority were in the north.

The shape of the land we see today was completely different then. If you lived on land, you could basically walk from England to mainland Europe. If you lived in Australia, the Great Barrier Reef did not exist; you could walk and hunt on dry land right out to where the reef is today.

Two major events happened at the same time to shape the earth we live on today. It is believed that the last major meteorite (around 3 km wide) to hit the earth collided with Northern Canada. This sparked the last major extinction event which caused the demise of the large mammals. Woolly mammoths, sabre-toothed tigers, giant bears, and huge ground sloths all perished.

We then had climate change. Simply put, it got warmer, but the change occurred over the next 1 ½ 000 years — unlike the rapid change we are seeing today. Over time, the great ice sheets started to melt and sea levels began to rise. They just kept getting higher and higher and higher until, just over 10 000 years ago, they reached the level they are at today. Sea levels rose by a huge 125 m (380 ft). This flooded an enormous expanse of the landmass as it was then, transforming our earth into what it is today.

FANTASTIC FACTS – ICE AND WATER

Imagine taking all the water off the earth, then flattening all the mountains and filling in the valleys to create a perfectly even-surfaced planet. If you then put all the water back, it would have a depth of 2.5 km (1.6 mi) completely encircling the earth.

All the land that we can see makes up 29% of the earth's surface, leaving 71% covered by oceans.

Sea level has risen 25 cm (9.8 in.) over the past 100 years and scientists expect this rate to increase to possibly over 1 m (3.2 ft) in the next 50 years.

It has been calculated that Antarctica has as much ice as the Atlantic Ocean has water.

The Arctic produces 5 000-45 000 icebergs annually. The number produced in the Antarctic regions is even greater.

The average Iceberg has a four-year lifespan.

WATER IN NUMBERS

How much water is there on earth and how much is in the oceans?

There is a total of 1 338 000 000 cubic kilometres of water in the ocean. 96.5% of all earth's water is salty, normally with the salt content of 34 parts per 1000. This means that 34 g of salt is dissolved in 1 kg of water—that's a lot of salt.

Coming into second place with just 1.76 % of earth's water is the ice and snow that makes up all the frozen water in the Polar Regions, mountains, and glaciers. It is a mere 24 364 000 cubic kilometres of fresh water. Alarmingly, this is decreasing every year.

All lakes, rivers, ground water, swamps and (let us not forget) the water in the atmosphere make up the rest. A total of 23 700 000 cubic kilometres of water.

So, give or take a few hundred cubic kilometres, the universe in the early period of the earth filled our planet with 1 386 000 000 cubic kilometres of water. That's one huge cosmic hosepipe!

FANTASTIC FACTS - TIDES

The tides do not come in and out every day. What happens is that the moon's gravity holds the water in the same position and the earth rotates, causing the

land to move in and out of the sea, rather than the sea moving up and down the shoreline.

The highest tides in the world are at the Bay of Fundy, which separates New Brunswick from Nova Scotia. At times, the difference between high and low tide is 16.3 m (63 ft) —taller than a three-story building.

On a global scale, the tide is a giant and incredibly long wave, moving around the globe at 725 km/h (450 mi/h). It has no beginning and no end.

UNDERSEA WONDERS

The largest waterfall on earth is actually under the sea.

The tallest waterfall on earth is Angel Falls in Venezuela, which has a drop of over 975 m (3 200 ft). That's a small stream compared to the Denmark Strait Cataract, which is an underwater waterfall between Greenland and Iceland, formed by the temperature difference in the water on either side of the strait. When the cold water from the east hits the warmer water from the west, something strange happens; the cold water is heavier than the warm, so it falls down the side of the warm water and underneath it. With a drop of 3500 m (11 500 ft), this makes it the largest cascade of water on earth. It is estimated that the flow of the waterfall is more than 3.4

million cubic metres (123 million cubic ft) per second, which is around 50 000 times that of Niagara Falls.

The Grand Canyon, a steep-sided canyon carved by the Colorado River in Arizona, is 1857 m (6093 ft) at its deepest point. However, the Zhemchug Canyon, located in the Bering Sea, has a vertical drop of 2590 m (8 500 ft)—about 730 m (2 400 ft) deeper than the Grand Canyon.

Humans have explored less than 5% of Earth's oceans.

Hydrothermal vents are splits in the sea floor that pump out sulphur compounds, supporting the only complex ecosystem known to run on chemicals rather than energy from the sun.

One study of a deep-sea community of animals living on and in the sea floor revealed 898 species in an area about half the size of a tennis court.

228 450 marine species have been identified, but there are very likely to be millions; scientists agree that over 80% of all life in the sea is undiscovered.

The freediving record by a male diver was 300 m (984 ft 3 in.) in Turku, Finland, on 3 July 2016.

A dolphin may be able to dive up to 300 m (1000 ft)

12 people have landed on the moon, yet only 3 people have been to the bottom of the Mariana Trench.

The deepest point on earth is located in the Mariana Trench, which lies in the western Pacific Ocean east of the Mariana Islands, hence its name. It is an underwater valley about 2550 km (1580 mi) long and 70 km (44 mi) at its widest point.

It's really deep —no it's really deep! —at around 10 984 m (36 037 ft) at a location called Challenger Deep (now that's a good name). If I took a really big knife and chopped Mount Everest off at ground level, then placed it in the Deep, there would be 2000 m (6560 ft) of water above the top of the mountain.

At the bottom of this trench the pressure on your body would be over 11,000 tons/m^2 —the same as having 50 jumbo jets on top of you. On land, pressure on your body is the weight of the atmosphere pushing down on you. Thankfully, gas does not weigh much; however, water is different. So, at the bottom of the trench you would have all the weight of the atmosphere plus the weight of the water acting on your body. All this equals a squashy mess!

On the first descent to the bottom of the trench, it was claimed that a 30 cm flatfish was observed, but in recent years this has been questioned. The deepest observed fish was in the Mariana Trench at a depth of 8145 m (26 722 ft), which was caught on video. One very sad thing to note is that crabs and amphipods sampled from the bottom of the trench were all found to have microplastic in their bodies. Even in the deepest parts of the ocean, humans can still find a way to pollute.

The Mid-Ocean Ridge is Earth's longest chain of mountains and is almost entirely beneath the oceans.

Where the earth's suboceanic tectonic plates push against each other, the crust rises to form a huge chain of mountains right around the globe. It is 65 000 km (40 000 mi) long. We know more about the surface of Venus and Mars than we do about this mountain chain.

90% of all the volcanic activity on the planet occurs in the mid-ocean ridge, and the largest known concentration of active volcanoes is in the South Pacific, containing over 1100 volcanoes in an area of 780 km^2 (480 sq mi).

The Pacific is wider than the moon.

It is estimated there's enough gold in the ocean for every person on earth to have 4 kg (9 lbs) each! At the time of writing that would be about $240 000 each.

WONDROUS WATER

Seawater, or what most people call salt water, is not just water and salt. It has, in fact, a very complex chemistry of over 138 different chemical molecules; it's just that salt (sodium chloride) is in the highest concentration.

Now water is a very special molecule, and at this point it's worth taking a look at it. Each molecule of water contains one oxygen atom and two hydrogen atoms, connected by two covalent bonds. It is this structure that makes the water molecule special.

The oxygen carries a negative charge and the hydrogen carries a positive charge. As opposite charges attract, the hydrogen and oxygen of different molecules basically join together via what is known as the hydrogen bond. They really don't like to be separated; they stick together. This can be seen when you look at the surface of water in a glass; it bends inwards because all the water molecules are stretching and holding onto each other, hence the name surface tension.

Water is also very unusual because of its unique chemical nature—when it freezes it expands. Therefore, when frozen, it weighs less. This makes ice float, which is very handy for life on earth. Because it floats, it basically forms a barrier between the atmosphere and the water. This ice barrier causes the temperature of the underlying water to remain relatively high and protect the life that exists beneath it. Ice is like a blanket, keeping everything warm.

OCEAN CURRENTS

Oceanic currents are a continuous movement of sea water in a predicted direction. They are mainly

generated by wind, breaking waves and differences in temperature or salt content (salinity).

Currents mainly flow horizontally, but vertical currents are hugely important and are known as upwellings or downwellings.

There are 34 major currents that circulate around the globe.

Ocean currents transport the heat from the equator towards the poles. They maintain the natural order and balance of the climate. Climate change will affect these currents and cause a global temperature shift.

The top 3 m (10 ft) of the ocean holds as much heat as the entire atmosphere.

So how do currents move?

It is a very, very complicated mechanism that pushes and pulls the water bodies around the globe. Let's try simplify it. Take a big coil of string to a park. You hold one end, and a friend starts to unwind the coil, while moving away. Soon they are 500 m in the distance, but you still have another 500 m of string by your side. You are both holding each end of the string tightly, and your friend pulls at the string. This forces your arm to move away in the direction of the pull, and string starts to move through your hand as your friend carries on pulling.

This is like a current. One end of the water body moves away and because water sticks together, it drags along all the water behind it, forming a moving conveyor belt. One end of the current pulls away whilst the other is dragged along, sucking in more water all the time.

The Gulf Stream is a current of warm water in the Atlantic Ocean. At a speed of 97 km (60 mi.) per day, the Gulf Stream moves over 100 times as much water as all the rivers on earth.

The density of sea water increases as it becomes colder, right down to its freezing point of -1.9°C. The average temperature of all ocean water is said to be between 2 and 3.5°C.

What do we mean by density and why is it important in the ocean? Imagine I had a bar of chocolate made up of 100 molecules, but I also had a bar of gold, exactly the same size as my chocolate bar, but containing 1000 molecules. We would say that the gold was denser than the chocolate as it had more particles within the same space.

Now, weight is a force generated by gravity pulling things towards the centre of the earth. Gravity holds onto the particles and pulls them downwards, producing a force called weight.

So the bar of gold will weigh more because it has more molecules for gravity to grab onto and pull downwards than the chocolate bar has.

As a rule, liquids get denser the colder they get because their molecules move closer together. Cold sea water is denser than warm sea water; therefore it is heavier. The heavier water will sink and the warmer, less dense water will float on top of it.

There are both surface currents and deep-water currents. As a surface current moves in one direction, the deep-water current underneath it moves in a different direction.

Why do we have different bodies of water moving next to, above or below each other? Water mixes, right?

Wrong! Take two glasses of water. Put a little bit of orange juice into one and a little bit of lemon juice into the other. What you have are essentially two glasses of water but with a completely different chemical makeup.

If we take the glass containing orange juice and heat it, we will have two different glasses of water with different chemical makeups, and they will also have different temperatures. If we could magically remove the glasses, we would find the two water bodies would not mix well. Perhaps they would mix a little where they met; however, they would remain separate because of their different chemical makeup. The warmer water would float on the surface of the cold water because of its lighter weight.

In the ocean we have bodies of water that differ in temperature and salt content; for this reason, they do not mix. The warmer water is pushed away from the equator until it reaches the cooler areas of the earth. There, the surface water cools; thus it gets heavier and starts to sink to the bottom of the sea, dragging more surface water with it. This becomes the much cooler deep-water current, which flows back towards the equator.

So we have a current originating on the equator; but what exactly moves it northwards or southwards? Essentially, the wind.

The prevailing wind, blowing constantly in one direction, causes the surface waters to be pushed in that direction, until they cool and begin to sink.

If the current in question hits the continental landmasses it can do one or two things, depending on different factors. The landmass can force the surface water down to the deep (a downwelling), or the landmass can force the deep water current up to the surface (an upwelling).

Warm water currents bring rain to the coastal areas because they supply moisture to the winds.

Where there are cold water currents, water evaporates at a slower rate, resulting in little or no rainfall in the coastal areas. The Patagonian Desert in South America is the result of the cold Falkland current.

Fog at sea is the result of cold and warm water masses meeting at the surface when there is little air movement. This causes huge amounts of evaporation without the moisture in the air going anywhere.

Thermocline is the name given to the area directly between two water masses of different temperature. The temperature can shift from as much as 20°C to 2°C in the distance of only 10 m (33 ft).

So, the thermocline is the temperature barrier between two water bodies. In the tropics it is a permanent feature within the ocean, yet in the temperate zones it develops during the spring and then breaks down in the autumn.

Life in the surface ocean produces an enormous amount of biological waste. This particulate waste has only one direction to go—downwards. There is such a large amount of this waste that when you are in the deep ocean looking out of a submersible, it is like looking out at a snowstorm. Only, the snow here is made of animal body parts and biological waste from the surface water.

This waste will eventually hit the sea floor, where some of it will be eaten. However, a great deal is broken down by bacteria and then released as nutrients back into the water.

This generates a situation where we have a nutrient-rich deep-water current separated via the thermocline from the nutrient depleted surface-water current.

Suddenly, the nutrient-laden water collides with a continental shelf and with its passage blocked by a sheer underwater cliff, it has only one way to go, which is upwards. We have an upwelling.

This vertical movement of water does not only create a cooling effect but supplies huge amounts of nutrients to replenish the surface waters, encouraging algae and plankton to grow. This natural fertilising process recycles all the nutrients, allowing life to develop and continue in the surface ocean.

Think of Florida, its warm climate and its coral reefs, bathing in the tropical Atlantic Ocean. Then draw a line across to the Pacific, where there are no coral reefs; in fact, the water is positively chilly due to the upwelling current.

WAVES OF ENERGY

Ocean waves are caused by wind moving across the water surface.

The tallest wave ever measured was 525 m (1720 ft) at Lituya Bay, Alaska.

The tallest wave recorded in the open ocean was 29 m (95 ft) which formed during a storm near Scotland.

Swells are rolling waves that travel long distances across the ocean. These are smooth waves rather than choppy wind waves. They are created by storms that are often thousands of kilometres away.

A swell is measured from the crest (top) to the trough (bottom) of the wave. So, a 10 m swell is the total height of the wave.

Now, waves are funny things; when you are standing on a beach watching them roll in, you would be forgiven for thinking it was the water that was moving. It's not.

The waves that you see at the seaside, 95% of the time, are generated by the wind. Wind is the movement of air, and these air particles blow over the water surface. When the gas of the atmosphere touches the liquid of the sea, friction occurs between the two surfaces. This friction causes a transfer of energy from the atmosphere to the water. It is this energy that creates a wave.

This is a hard concept to grasp: when looking at waves, you are not looking at the movement of water. Instead, you are actually looking at the movement of energy contained in the water.

If you looked at a water particle moving, you would find that it moves up and down at right angles to the wave. The water itself does not travel with the wave; it just moves up and down. The effect that we see— water moving with the waves—is instead the result of the energy contained in the waves, which is moving. This is very hard to visualise, but it is a law of physics: all waves are the movement of energy.

TERRIBLE TSUNAMIS

In 2004 the Indian Ocean tsunami was caused by an earthquake with the energy of 23 000 atomic bombs. Multiple tsunamis hit 11 countries. The final death toll was 283 000.

About 80% of tsunamis happen within the Pacific Ocean's 'Ring of Fire' due to its high underwater volcanic activity.

Tsunamis can travel at speeds of about 800 km/h (500 mi/h.) — the speed of a jet plane.

SUBMARINE SOUND

The speed of sound in water is 1435 m/s (4708 ft/sec), nearly five times faster than the speed of sound in air.

Sound is another fantastic wave in the oceans; animals use it to communicate, hunt and avoid predators. Humans use it to map the ocean floor and to find fish and shipwrecks.

Sound is not like our surface waves. Instead of the partials moving up and down, this wave pushes and pulls. A speaker in your headphones, sound bar or phone vibrates to cause sound.

As the speaker pushes forward, it compresses the air directly in front of it, and as the speaker moves backwards, it stretches the air particles backwards. Then it pushes forward again and compresses again and so on and so on. This happens many thousands of times a second and a wave of sound energy is produced. We have a wave that is a long continuous line of areas of particles compressed and areas of particles stretched; physics calls them compression to and rarefactions—it's a longitudinal wave.

Just like our water wave, it is the energy in the wave that travels, not the particles. Particles move backwards and forwards, and in doing so they hit the particles in front of them. It is this collision that causes the energy to be transferred from one particle to another to move along the wave.

The particles in gases are very far apart compared to water particles, which are close together. Therefore, sound travels faster in water than air.

FANTASTIC FACTS – SUBMARINE SOUND

A dolphin can pick up a sound from over 24 km (15 mi) away.

The sperm whale produces the loudest sound made by any animal on the planet. At 230 decibels, if you were to stand by its head when it made this click, it would kill you.

MARITIME MISADVENTURE

A very sad aspect of the oceans is the massive loss of life that has occurred every year since we started to sail.

An estimated 3 million shipwrecks are spread across ocean floors around the planet.

The oldest discovered boat in the world is the 3m-long Pesse canoe, constructed around 10 000 years ago. More substantial crafts existed earlier. A rock carving in Azerbaijan dating from 12 000 years ago shows a boat manned by about 20 sailors, all holding paddles.

We can be certain that shipwrecks have occurred for at least the last 12 000 years. It will hardly be questionable then, when we look at the oceans, that they will contain more artefacts than all the world's museums put together.

The greatest single loss of life made the Titanic look like a small rowing boat; not surprisingly, it occurred during World War II.

The Wilhelm Gustloff, a German military transport ship, was sunk on January 30, 1945, by a Russian submarine S-13. One estimate is that 9400 people, including around 5000 children, died. German forces were able to rescue 1,250 survivors. Just 11 days after the S-13 sank another German ship, General von Steuben, killing about 3000 people. I wonder if this is why the number 13 is termed unlucky.

More than 90% of the trade between countries is carried by ships and half of communications between nations use underwater cables.

THE PLANKTON

Many people think plankton are microscopic and most are. The word plankton actually derives from the Greek language, meaning drifting life. Any living organisms that cannot actively swim against the current are categorised as plankton.

Science has separated plankton according to their size distribution.

Picoplankton: Less than 2 micrometres
Nanoplankton: Between 2 and 20 micrometres
Microplankton: From 20 to 200 micrometres
Mesoplankton: Between 0.2 to 20 millimetres
Macroplankton: From 2 to 20 centimetres
Megaplankton: Over 20 centimetres

A mouthful of seawater may contain millions of bacterial cells, hundreds of thousands of phytoplankton and tens of thousands of zooplankton. How many have you eaten?

The 2 major groups of plankton are phytoplankton (plants) and zooplankton (animals).

FABULOUS PHYTOPLANKTON

Phytoplankton are microscopic algae. Like plants on land, phytoplankton form the basis of life in the oceans. They are the primary producers, manufacturing their own food using energy from the sun, carbon dioxide from the air, as well as water to produce glucose and oxygen.

They also need a wide range of nutrients, which they absorb from sea water through their cell membranes. Because of this, they only exist in large concentrations within nutrient-rich water. Just like the plants in your garden, which need feeding, phytoplankton need the fertiliser within the sea. This is why it is so important that the nutrient-rich deep ocean currents can upwell to fertilise the surface waters.

Tropical waters are termed nutrient-deficient and therefore cannot contain a high concentration of phytoplankton. This simple fact explains why tropical water is so crystal clear.

Diatoms are a type of phytoplankton with calcareous or silicon-based skeletons. The white cliffs of Dover and all the limestone in the world is made mainly from diatoms and other hard-shelled phytoplankton that existed hundreds of millions of years ago. That's a lot of plankton!

Also hundreds of millions of years ago, large numbers of plankton bodies piled up on the sea floor and were buried. As they were buried, they were pushed deeper into the earth's crust. Over millions of years, extreme pressure and the earth's heat changed these plankton bodies into natural gas.

A high proportion of diatoms living today are found in fossil records from over 100 million years ago—incredibly old species indeed.

BLOOMING ALGAE

Phytoplankton need light and carbon dioxide from the air. This means that they have a very low tolerance to environmental change. Most species cannot tolerate temperatures over 20°C. They need nutrient-rich, well-circulating water and light. Because of this, we find the major populations occur within the temperate zones, polar regions and areas of upwelling in the tropical areas of the globe.

In springtime, when the winter ice sheets start to melt, the temperate waters begin to warm and sunlight hours gradually increase, it's happy days for the phytoplankton. All winter, the surface waters have had a continuous and steady replenishment of nutrients. Temperature and sunlight are now increasing to create the perfect storm for phytoplankton growth—and do they grow!

There it is: a happy little single-celled alga bobbing around in the surface ocean. It's been dormant for a few months, taking a lovely sleep, and thankfully, through some biological miracle, it has not come across anything that wanted to eat it.

The sea is warming and the alga begins to wake up. Soon it is absorbing nutrients and carbon dioxide, producing its own food and beginning to grow. It doesn't really grow in size, but instead, it simply makes a new cell—it divides in two. After the first 20 divisions from one cell, it has produced a total of 1048.576 cells. It had 100 friends at the beginning and after 20 divisions they produced a total of 104 857 600 phytoplankton cells. If we then take the last figure and look at another 20 divisions, it is clear that the algae are suddenly increasing their biomass rapidly—they are starting to bloom.

An algal bloom is defined as an increase in biomass of one or more species of phytoplankton, exceeding 1 million cells per litre of water.

Phytoplankton growth rates depend on many factors; some species divide every 2- 4 hours in a 24-hour period while others only divide in daylight.

Blooms are so intense that the huge number of cells turn the water the colour of the algae. They can turn the sea deep green, orangey green and even a deep red, termed a red tide.

Bioluminescence is the name given to the process where light is produced by a living organism. The chemical responsible for this is called luciferin. When this molecule reacts with oxygen, it produces light that is normally iridescent blue. Some species of phytoplankton are capable of this.

Truly one of the most amazing sights that anyone can behold is the bioluminescence of an algal bloom, only seen at night on calm seas. If you are lucky enough to be on the shore or in a boat at the right time and in the right place, you will witness a spectacle you will never forget.

 As the waves lap onto the shore, or the water is disturbed by a thrown stone or by the wake of the boat you're sitting in, a biological chain reaction occurs within the algal cell. This produces a fluorescent blue light, turning any disturbed area of water from a dark black shadow to an intense beam of bright blue light. It has to be seen to be believed.

Space satellites have been used to follow the movement of the blooms; this shows the speed and direction of the surface current where the bloom is occurring.

Algal blooms are responsible for a huge loss of aquatic life, and have caused countless trips to the toilet and even death in our human population. (More of that later.)

HARMFUL ALGAL BLOOMS

HAB is the abbreviation for harmful algal bloom. A mass mortality event of 107 bottlenose dolphins occurred along the Florida coast in the spring of 2004 due to their feeding on menhaden (a fish species) that had been exposed to an HAB.

Not all species of phytoplankton are harmful when blooming.

Imagine a scenario where we have a nice algal bloom occurring in the coastal waters, and the mixing of the surface water causes the depth of the bloom to deepen. So not only will the bloom extend over the surface, but it will also move deeper and deeper into the water column. A situation develops where seawater has been transformed into a thick soup of algal biomass. The many algal cells produce various harmful waste products, an immense amount of mucus and sometimes release toxic gases into the atmosphere. All this can be hazardous to aquatic life—suddenly we have a harmful algal bloom.

The immense density of the cells is deadly to any fish swimming through them, as the mucus sticks to their gills and clogs them up, resulting in rapid death. Quite often, dead fish can be found on the seashore over a range of many kilometres.

The algal bloom won't last forever; if it dies in the shallow waters, dead cells will sink to the seafloor and smother the sediment. This layer becomes so thick that it forms a barrier over anything living in the sand. An explosion of bacteria results, which feeds on and breaks down the dead cells, while also consuming all the oxygen in the sediment. Soon, a thick black mud develops, stopping any oxygen in the water from entering the sand. Any and all life within that sand is doomed.

HAB's s are subject to a high degree of scientific study. We need to know where they're occurring, and when. Scientists are collecting years and years of data to see if these are cyclic events. The problem for the scientists is that only in recent times, using satellites, have we been able to track them on a global basis.

What are we looking at? Blooms that occur naturally every year are less productive in some years and more productive in others. Are these cyclic over time—

occurring every 10, 20 or 30 years? One thing we do know is that in areas of large runoff into the sea (in other words, rivers containing a high degree of fertiliser from farmlands), we have a higher occurrence of algal blooms and also a higher intensity of individual blooms. This provides yet another example of human activities disrupting the natural balance of the sea.

FANTASTIC FACTS – ALGAL BLOOMS

One bloom in northern France was caused by fertiliser runoff. The algae produced a toxic gas as a by-product, and when the wind changed direction and moved towards land, this killed farm animals and was directly linked to one human fatality.

The largest algal bloom on record occurred in 1991 in the USA and stretched over 1000 km (620 mi).

In 2010, dissolved iron in the ash from a volcano caused a huge bloom in the north Atlantic.

CARBON SINKS AND CHALKY SHELLS

2.6 billion tonnes of carbon from human activities are absorbed by the ocean every year.

If we have water which has 100 molecules of carbon dioxide in it and the atmosphere around the water also contains 100 molecules of carbon dioxide, then the system is in balance; the carbon dioxide will stay where it is. However, if the water loses some of that carbon dioxide, let's say 50 molecules, we have a situation where there is more carbon dioxide in the atmosphere than in the water. Through the natural process of diffusion, some of the more highly concentrated carbon dioxide from the atmosphere will move into the water, where it is less concentrated.

This is very important when we consider rising carbon dioxide levels in the atmosphere. The oceans are brilliant at absorbing carbon dioxide. Although the oceans absorb this gas, it is the phytoplankton that are taking it away. Many species have calcareous shells, requiring carbon to build— that carbon is absorbed from the water. Our phytoplankton absorb this carbon, for building their cells and shells; then they die and sink, taking the carbon with them. As the phytoplankton remove carbon from the water, the water absorbs more carbon dioxide from the atmosphere, helping to reduce the impact of global warming.

If too much carbon dioxide is absorbed by the sea and there are no phytoplankton to remove it, this causes the sea to become more acidic—it produces carbonic acid. In tropical areas where there is a low concentration of phytoplankton, we are seeing a problem where seas are becoming more acidic. This is contributing to the coral reef decline.

PLANKTON OR FISH?

When is a member of the plankton not plankton? When it is a sunfish or *Mola mola*. What an amazing fish! Here are a few facts about it.

The sunfish is huge and was classified for years as a type of plankton, because it seemed to drift with the current rather than swim. However, scientific tracking has shown that sunfish swim at speeds similar to those of other large fish.

Its average size is 3.3 m (10 ft) long and it weighs 100 kg (2200 lbs), but the biggest can grow up to 2270 kg (5000 lbs) making them the world's largest bony fish.

The sunfish can produce 300 000 000 eggs, the most by any animal alive.

These fish were accepted by Japanese shoguns in the 1600's as payment for taxes.

The name *Mola mola* comes from the Latin word for "millstone." Well, it does look like one.

WHO'S WHO IN THE ZOO?

The next step up in the aquatic food chain is the zooplankton. We have two separate types of zooplankton: the permanent members are those species whose whole lives are spent as plankton and the transient members are those who only spend some time as plankton before they grow up and drop out of this category.

Zooplankton can be the eggs and larval stages of different species such as fish, crabs, starfish, shellfish and so on—the list is endless. Apart from these, the zooplankton are also miniature predators consuming phytoplankton and other zooplankton, with larger species even eating fish.

FANTASTIC ZOOPLANKTON FACTS

Zooplankton reproduction can be rapid; some species have recorded a 30% population increase over 24 hours.

If zooplankton were left alone and not eaten, their dead bodies would cover the entire ocean floor a metre deep in 130 days.

The biggest known species of plankton is the lion's mane jellyfish. The largest recorded specimen had a body that was 3 m (9 ft) across and tentacles that were 36 m (120 ft) long.

A group of herring is called a seige. A group of jelly fish is called a smack.

Shrimp can only swim backwards.

Ángel León operates a Michelin 2-star restaurant in Cadiz in the South of Spain, where plankton occupies a place of honour on the menu.

Sheldon J Plankton, a zooplankton with one eye and 2 long antennas and an evil character in the carton SpongeBob, is a very good depiction of a common species of zooplankton called a copepod. They do look like him.

In 2008, 3 scientists (who won the Nobel Prize for chemistry) were able to extract and develop a green fluorescent protein, a substance that glows intensely under ultraviolet light, from a species of zooplankton—a jellyfish. Injecting this into human cells, doctors can track the way cancer tumours spread or how brain cells develop.

The greatest migration on earth happens every night. It is the largest movement of biomass on earth. The great wildebeest migration in Africa or the huge amount of salmon that migrate on a yearly basis are nothing compared to the nightly zooplankton migration.

During the Second World War, sonar operators in the US Navy started to report very strange readings. Their sonar readings suggested that the whole of the ocean floor was rising at night, then as dawn approached, the sea floor started to fall. This was obviously a great international concern.

They got the biologists to investigate and soon realised that the sonar wave was bouncing off shoals of animals as they moved upwards through the water column. It was not a secret weapon and it wasn't the sea floor moving, but it was, in fact, the movement of zooplankton. Their shoals were so dense that they disrupted the sound wave and bounced it back to the sonar receiver, giving a false sea floor reading. The greatest migration on earth had been discovered—but more of that later.

INCREDIBLE KRILL

The most famous member of the zooplankton is the krill, a shrimp with over 85 species.

In the Southern Ocean alone, one species, the Antarctic krill, *Euphausia superba*, makes up an estimated biomass of 400 or 500 million tonnes. It is estimated that the total weight of all krill species is twice that of humans.

One of the oldest living species on earth is the krill *Bentheuphausia ampblyops*, which lives in the deep waters below 1000 m (3330 ft).

Whales eat about 4% of their body weight in krill daily during a 4-6-month summer feeding frenzy in order to build up fat reserves before their breeding migrations.

Krill can live up to 10 years—not bad for a shrimp—but most don't make it!

Krill oil is considered a health food since it consists of various essential nutrients. Krill oils are also used to treat high blood pressure, stroke, cancer, osteoarthritis and depression.

MINI-MONSTERS and GREAT JELLIES

The permanent members of the zooplankton range from the microscopic to the huge. Unlike phytoplankton, which require light to live, these can be found in the deepest parts of the ocean. Many species spend their whole lives there, in an area that has never seen light.

The microscopic zooplankton are indeed mini monsters; if they were large, they would be the stuff of horror films.

Zooplankton are bigger in size and, as such, weigh more than the phytoplankton. The problem is, they are more at risk of sinking. Moving and swimming, even with the current, uses up vital energy— energy the plankton don't want to waste. Both phytoplankton and zooplankton have come up with elaborate ways to slow down their sinking rates.

Imagine you are floating in the sea (well, slowly sinking), so you have to tread water to stay afloat. You come across a plank of wood 8 times the length of your body, grab hold of it and now you can float.

The same thing has happened to many species. They grow long appendages: spikes and arms which extend out into the water to slow down the sinking problem. Some carry pockets of gas to aid their buoyancy, some have gone one step further by filling up a balloon with gas and spend all their lives floating on the surface. Others grow in colonies, forming long chains of individual animals or plant cells, these all act together to reduce the sinking effect.

Zooplankton can also bloom. During the 1990's, details first began to appear in the scientific literature of increasing instances of jellyfish blooms. Jellies are a nuisance and a pest to humans, with many possessing extremely painful and even deadly stings.

Currently we are experiencing huge explosions of jellyfish populations occurring in waters around the world. It has been estimated that over 120 tonnes of wet jelly weight can be achieved by one bloom.

If this occurs near a fish farm, the jellies can clog nets or lodge within gills, starving valuable fish of oxygen. It's a terrible death for the trapped fish and something that can cost over $50 000 per fish pen.

A bloom can seriously threaten coastal power stations. The swarm can block underwater cooling systems resulting in shutdowns. The Torness nuclear power station, situated on the coast near Edinburgh, Scotland, was forced to shut down for several days in 2011 due to an approaching swarm of jellies.

Zooplankton is also the classification for the larval forms and eggs of hundreds of thousands of marine organisms. Most species will lay eggs directly into the water, shedding eggs and sperm in a mating frenzy, hoping sperm will meet egg and fertilize it. Others lay eggs under rocks, seaweeds and other structures where larvae can hatch and be welcomed into the zooplankton community for a brief period of time.

A brief period of time is an understatement. Some eggs and larvae only live for a matter of moments, existing only to feed another beast. If 2 out of 1 million eggs make it out of their planktonic stages and into their adult body forms — that's a good result.

As plankton, many animals do not look like their adult forms. Instead, they grow and transform through different stages and body shapes until they reach their adult shapes. Most crabs will develop through as many as 8 or 9 zoea larval stages during their planktonic lifespan. The final zoea larva then moults into a free-swimming larva called a megalopa—a transitional phase between the planktonic zoea and the adult.

FANTASTIC FACTS – CURIOUS CREATURES

Only 2 fossil crab larval stages have ever been found, one is 110 million and the oldest is 150 million years old.

Many species lay their eggs to coincide with algal blooms so that when the eggs hatch, the larvae will have a food source.

Horseshoe crabs have existed in essentially the same form for the past 135 million years.

Left to their own devices, pearls grow naturally only once in every 20 000 oysters.

LOCOMOTION and MIGRATION

The first thing we need to understand here is the difference between movement, locomotion, and migration. Movement involves moving your hand, your head or spinning on the spot. Locomotion is the movement of your entire body from one place to another, like a shark swimming from the reef to the sandy shore. Animal migration is defined as the regular, usually seasonal, movement of all or part of an animal population to and from a given area. In this chapter we will look at locomotion and migration within the marine environment.

The gold medal for the slowest animal in the sea and probably the world (one or more could still be discovered) is awarded to…the sea anemone. With over 1000 species, 90% of these creatures are all putting in their claims for the title. We don't know which one is the slowest because there are so many different factors affecting their speed and most importantly, we have not placed a time-lapse camera in front of them all to calculate it. However, it is known that many species do not exceed 0.0001 km/h, or to put this into context, it would take one anemone 41 days without rest to travel a distance of 1 km or one year to go 9 km.

FANTASTIC FACTS – THE SLOW AND THE SPEEDY

Swordfish and marlin are considered the fastest fish in the seas, topping speeds of over 120 km/h (75 mi/h).

It would take an anemone 13.5 years to travel the same distance (without sleep) that a swordfish could swim in one hour.

The dwarf seahorse is the slowest-swimming fish with a top speed of 1.5 m (5 ft) per hour. So it will take 3.5 days for this seahorse to cover the same distance a swordfish would travel in an hour.

Bluefin tuna can travel up to 90 km/h (55 mi/h) over long periods. Highly prized for sushi in Japan, bluefins are also among the most valuable fish in the world and sell for over $20 000 each.

Do penguins swim or fly underwater? It is a debate! But they do it at up to 40 km/h (25 mi/h).

At night, a dolphin sleeps just below the sea surface, shutting half its brain down and leaving one eye open. This allows it to sleep and rise to the surface to breathe, whilst constantly swimming.

Why do dolphins jump out of the water? Is it to conserve energy, to see what's going on or just for fun? No one really knows.

The fastest dolphin is the common dolphin, reaching speeds of 60 km/h (37 mi/h). Two species share second place, both achieving speeds of 55 km/h (34 mi/h). These are the killer whale, being the largest dolphin at 8 m (26 ft) long and weighing up to 5.5 tonnes (857 st) and the 2 m (7 ft)-long Dall's porpoise, weighing in at 200 kg (440 lb).

The fin whale is nicknamed the 'greyhound of the sea' as it is the fastest whale, attaining short bursts of up to 46 km/h (29 mi/h).

When a dolphin is injured, it lets out a call for help and other dolphins quickly come to its aid, making sure it is able to breathe. Its call for help can also be a call to dinner for any large predator!

Most sharks do not consider humans as food, but on rare occasions, a mako shark has attacked a human. It is the fastest-swimming shark, so you have no chance of out-swimming it at speeds of 80 km/h (50 mi/h).

BREATHING UNDERWATER

All animals need a constant supply of glucose and oxygen to their cells where, via a chemical reaction, the two mix and release energy. It is this energy which keeps animals alive and allows them to move.

So how do fish breathe? Or to be correct, how do they pass water over their gills to extract oxygen from it and then transfer the oxygen into their blood? The three ways in which this occurs have great implications on how species move through the water.

Buccal pumping is the most common method employed by fish. Cheek muscles are named buccal muscles, and a buccal cavity is the interior of the mouth. Fish pump water into the buccal cavity and force it out over the gills where gas exchange occurs, then oxygen is transported around the body.

Ram ventilation is when fish simply open their mouths whilst swimming, forcing water through their mouths and over their gills. Ram ventilation is over 5 times more efficient in obtaining oxygen than buccal pumping. Many species switch between the two; various sharks, including sand tigers, use ram ventilation when swimming in short fast spurts—often when hunting. This allows an increased supply of oxygen to the muscles, giving them more energy, just when the fish needs it.

A few fish and sharks require so much oxygen in their muscles that if they stopped swimming, they would die in a few minutes. Here we enter the strange world of animals that never sleep: the obligate ram ventilators. These include fish such as the tunas and swordfish, with shark species being great whites, mako's, basking, hammerhead, thresher and whale sharks. They don't sleep as we do, but instead they enter a restful state where they slow down and cruise whilst shutting down certain body functions to give their muscles a rest.

However, they have to keep swimming 24 hours a day for as long as they live.

Why do over 99% of animals in the sea move forwards? The location of their eyes or light organs ensure that forward-facing is the norm. Try walking backwards—even when you know nothing is in the way, it is still unnerving. In the ocean, to see is to eat or be eaten, so evolution has progressed to encourage movement in the direction of sight. However, a few animals do swim backwards.

FANTASTIC FACTS – MIND-BOGGLING MOVEMENT

Triggerfish are unique because most of the time they swim only using their top and bottom fins and not their tails. This means they can hover, swim upside-down and even swim backwards.

The Scallop, a shellfish, can dart backwards by opening its shell and snapping it shut with amazing speed. This results in the shellfish being propelled through the water for up to 1 metre—great if you're going to be eaten.

Some shrimp species have paddles on the undersides of their bodies and can quickly dart backwards. Krill can also do this, but with the added advantage of shedding their skins at the same time.

All a hungry fish gets is the empty shell. This is literally jumping out of your skin!

INCREDIBLE CEPHALOPODS

Cephalopods such as the octopus, squid and nautilus swim by jet propulsion. They ingest water and force it out of a small opening called a syphon. It's like squeezing the end of a hosepipe so the water is forced out at a much higher rate than it came in. The syphon is also moveable, allowing these creatures to move in a particular direction.

Cephalopods are very close relatives of shellfish and sea slugs.

There are over 800 species of cephalopods with over 11 000 species known only from fossil records dating back over 500 million years.

Cephalopods are the Einstein's of the invertebrate world. Their relatively huge brains allow them to perform many tasks other invertebrates cannot, including communication via changing the colour of their skins. They use this colour change to find a mate, for camouflage and warning off others— even though most are colour blind!

When is bigger actually smaller? When you are a colossal squid. This beast is much bigger than the famed giant squid, yet not as long.

We can only estimate, but specimens collected to date show the colossal squid weighing in at 490 kg (1100 lbs) with a length of 10 m (33 ft), while the giant squid weighs 275 kg (600 lbs) but measures over 14 m (46 ft) long.

Coming close third is the pacific octopus with the largest being 136 kg (300 lb) in weight with an arm span of 9.8 m (32 ft).

SLIDERS AND GLIDERS

As mentioned, shellfish are close relatives to the above; some attach themselves to a surface and never move from that place, yet many can glide over surfaces at a relatively fast pace using only a foot. Muscles and slime are the major players here. Think of ripples on a water surface—this is what the underside of the muscular foot looks like. Hundreds of ripples over a slimy surface force the shellfish along over rocks or sand, much like a plank of wood rolling over the waves onto the shore.

One of the largest mobile shellfish is the queen conch, growing up to 30 cm (1 ft) in length. Its individual speed is not known, but this conch can have a territory of up to 15 acres in size. That's one very large kingdom for a 1-foot animal when you consider an acre is 43 560 sq ft—that's 653 400 sq ft per conch!

EXTRAORDINARY ECHINODERMS

Not all animals that glide over surfaces within the wonderful oceans use muscles to move. Enter the amazing arena of the echinoderms: starfish, sea urchins, brittle stars and feather stars. This is the largest grouping of completely marine animals found in every location in the world, with over 7000 species recorded and probably hundreds more waiting to be found.

Think of a typical starfish with five arms, yet these are not used to walk. Underneath each arm are hundreds of tube feet, all working independently to attach and move the animal over a surface.

Imagine a long hose pipe with hundreds of taps along its length. Each tap has a balloon connected to it. The hose fills with water and one tap is turned on, inflating the balloon; the tap turns off and the one next to it inflates, and so on. Then the first tap is opened, and the balloon deflates—you get the picture. This is the water vascular system of the echinoderms; each balloon is a tube foot inflating, moving and then deflating as the animal glides over the surface. One animal—hundreds of tube feet and one long pipe filled with water; how evolution can create such wonders is truly amazing.

The largest starfish is the sunflower sea star, growing to 1 m (3.3 ft) across and possessing up to 40 arms. Using its 15 000 tube feet, it is the fastest starfish, travelling at 60 m/h (200 ft/h).

Sea Stars do not have brains or blood; instead of blood they use the water vascular system to transport nutrients around their bodies. Having no centralised brain can have its advantages when it comes to reproduction— more of that later!

GOING WITH THE FLOW

One thing all animals don't like doing is wasting energy; they prefer conserving it. Think of an eagle soaring in the sky—it's not flapping its wings—it's just floating, using its wings as support and moving while using the least amount of energy possible. Beasts within the sea are just the same, particularly the drifting life called plankton.

Spiders can be found high up in the atmosphere; they secrete a thread that gets caught by the wind and off they go. A mussel larva will do the same; the thread an adult uses to attach itself to a rock stretches into the water and carries off its baby to find new areas to colonise. This is the major reason that oil rigs are mussel-covered within weeks.

If planktonic algae and animals were bigger, their shells would be highly prized, as they are more diverse, weird and beautiful than any other shells in existence. Their designs are wondrous; some are sail-like or curly and long, others have huge spikes— all with one thing in mind— to catch the current and drift. It's locomotion without energy (well, nearly).

UNPACKING MIGRATION

On occasion, hundreds and sometimes thousands of starfish have been swept onto a sandy shore, leaving their beached bodies exposed to the air, where they die. Why this happens is a mystery. Some suggest that if a population in deeper waters used up its food supply, all those starfish might migrate to shallow waters to forage for food. There, they could be caught by the outgoing tide or by stormy waves throwing them up the beach.

Migration is a very, very complicated subject indeed. It is recognised that there are two main forms of migration, which are each further split into four other subgroups. The two main forms are obligate migration, where animals must migrate for survival and facultative migration, where individuals of species choose whether or not to migrate, often depending on the availability of resources.

All of them are triggered by environmental conditions, which change on a daily to seasonal basis, telling animals it's time to move. The two main reasons marine animals migrate are to breed and find food, but there are many minor reasons. With climate change, migrational behaviour is changing.

More than 75% of earth's marine life is migrating to different places and altering its natural breeding and feeding because of warmer waters.

Ocean species are migrating in response to climate change over 8 times faster than species living on land.

Some marine species have migrated over 950 km (600 mi) from where they existed in great numbers in just a few decades. Whole populations within an ecosystem are shifting. So, it's not just the horrific thought of entire species becoming extinct due to climate change but also whole populations beginning to move into new areas. What effect this will have on the species living in those areas is unknown.

One of three things will happen in any habitat: 1) the invading species will be fought back; 2) the invading species will live alongside the existing species but in fewer numbers; or 3) the invading species will decimate the population of the original species. That is a very, very simplified scenario of what is to come. It is an area of great concern as we are already facing a situation where this is happening. Enter the realm of invasive species…

An invasive species is an introduced organism that negatively alters its new environment.

An example is the lionfish. It's a perfect predator, being quite large with a huge mouth and venom-filled fins, which can be extended to make the fish look larger than it is (this discourages predators from eating it). This beast is naturally found all over the Indo-Pacific region.

Aquarium keepers in Florida, USA, let their lionfish out when they got too big for their tanks. Suddenly those fish found themselves in warm waters, with loads of smaller fish to eat and no natural predators. Their population has bloomed over the past 20 years, and they are, quite simply, gobbling up every smaller fish they can find—altering the natural balance of the ecosystem.

This is just one example of literally hundreds of invasive species (animal and plants) that are now at work altering the natural distribution of life on a global basis!

MARATHON MIGRATIONS

Scientists often attach an electronic tag to an animal which sends out signals about its location, measured not only in distance, but also in depth. It is easier to tag large animals than faster-moving small fish, so much more is known about marine mammals and turtles on an individual basis. However, due to the large numbers of fish in any one population, fish migration patterns are also relatively well known. As always, with marine science there is so much more to be discovered.

There are over 80 species of whales and all migrate. Whales mostly migrate to the colder poles during the summer for feeding and then to warmer waters to breed and give birth. Within marine mammals, the longest round-trip migration gold medal is held by the North Pacific grey whale at a confirmed 20 000 km (12 400 mi) round trip.

The longest swim recorded was by one leatherback turtle travelling from Indonesia to Oregon, USA and back again to lay her eggs. This is a mind-blowing total of 41 000 km (25 500 mi)—Wow!

The northern elephant seal migration takes place twice a year; each time the seal travels 10 500 km (6 500 mi), so when we see films of them piled up on beaches, I think they have earned a rest.

Some killer whales migrate for an odd reason—to change their skins. They travel the 10 000 km (6 200 mi) distance from Antarctica to warmer waters so that when they do shed their skins, they remain warm.

One fish that has been tagged was a marlin swordfish, swimming 15 000 km (9 300 mi).

Marine animals migrate, covering vast distances and, most of the time, arriving at exactly the right destinations at the right time— all without a route planner or satnav. How are they so specific? Finding the beach they hatched on, the river they were born in, the area with the right amount of food at the perfect time, meeting a member of the opposite sex at the right location to breed—I could go on and on!

Different species use a variety of cues at different times of the day and at various locations throughout their journeys; it's an ever-changing picture. They utilise the position of the sun and moon in the sky, the earth's magnetic field, gravity, temperature and (very importantly) smell. It's really complicated for us humans to work out how they do it, yet for the animals themselves—well, they just do it. It's instinct; it's what they do—no big deal!

FROM OCEANS TO RIVERS (AND BACK)

One of the most amazing facts is that a few species don't only migrate over huge distances, but they also move from fresh water to salt water and back again. From a biological viewpoint this is amazing; for over 99.5% of all aquatic species, this would result in a very quick death. The lack of salt, or the presence of salt would either shrink their body cells or cause them to swell so much they would burst. Here we encounter the bizarre domain of the euryhaline organisms, with the two best-known species being the salmon and Northern European eel.

JOURNEY OF A SALMON

Salmon migration is the subject that most scientific research has focused on, resulting in a huge knowledge bank for all populations. Why? Salmon is a valuable resource. Prior to aquaculture, only wild salmon were caught for food. There was also the tourism effect, where anglers would pay handsomely to travel to remote areas for the chance to catch the fish of a lifetime.

If something is worth a lot on money, then cash is available for research.

There are seven species of true salmon, which have a lifespan of between 3 and 13 years— I know which one I would be, given a choice.

Let's follow a mature salmon starting the long trip from its ocean feeding grounds to its home stream at the top reaches of a river.

Using sensors, its instincts come into play and it follows the earth's magnetic field back to the estuary. Here, it stops feeding because there is only one thing on its mind and that is to find a mate. (So, how do anglers catch a fish that is not interested in feeding? They present a food item right under the fish's nose, instinct causes the fish to strike, and it's hooked.)

When the fish enters the river, a most unusual sense comes to the fore: smell. This is not smell as we would usually think of it; instead, the salmon smells the chemicals in the water. It is looking for the unique aroma of a distinct area of water in that river or a stream that feeds the river. Unlike instinct, this is based on a memory that was imprinted into the fish when it hatched. If the salmon swims past the spot, it will turn around and go on the search until it has returned home. It then finds a mate and passes on its genetic information before it dies.

THE NORTHERN EUROPEAN EEL

Still not fully understood, the eel is an amazing beast with an interesting but fragile life history. This one fish illustrates how climate change can cause the extinction of a species.

These slimy, slippery creatures can live up to 85 years, reaching a length of 1.5 m (5 ft) and weighing over 10 kg (22 lbs).

Not only do they move from salt to fresh water, but when in lakes and rivers, they can slither through wet grass on land for over 500 m to find another water body.

Eels are critically endangered—records show that over the last 35 years, eels arriving in Europe have declined by over 90%. Changing techniques in farming and habitat loss are considered to be the major cause of this decline.

Now, to one mind-blowing life history that truly defines what migration is about and highlights the fragile earth we all share.

We shall start where life begins for this animal as a fertilised egg in the Sargasso Sea of tropical Bermuda—named so for the huge amount of floating sargassum seaweed there. This is an area where 4 major oceanic currents meet, causing the water to move in a circular motion, which leaves the floating seaweed trapped by the currents.

However, our fertilised egg is so small that it is not trapped; it exits the sea and starts to drift east with the Gulf Stream. Two years later, it reaches the shores of Northern Europe. What was once an egg has transformed from a leaf-shaped larval stage to a transparent (predators cannot see it) mini-eel as it reaches the river estuary. Only two animals for every 1000 eggs live to tell the tale.

Here, the eel swims inland and changes to a dark brown colour, swimming upstream and slithering over land to find ponds and streams where it grows for at least 20 years. Then it starts to migrate back again, but unfortunately, as for many migrating fish, this is the beginning of the end.

When an eel reaches salt water, its gut dissolves and feeding becomes a thing of the past. Not distracted by finding food, its only thought is to make the journey as quickly as possible. From the English Channel to its destination takes between 165 and 175 days.

How they do it is unknown; the major theory is that eels sense the earth's magnetic field and follow it, but other factors will also contribute, such as smell, the position of the sun and many yet-unidentified factors.

One thing that is known, is that if global warming continues and a critical sea temperature is reached, the Gulf Stream will change direction. When this happens, the eels' eggs and babies will drift to a completely different destination where no rivers can be found.

The Sargassum Sea may simply not exist anymore, as the 4 currents making it may shift location and the eels' breeding ground will be lost. This will result in the extinction of the species—millions of years of evolution knocked out by a few years of human activities.

FEEDING

When it comes to fine dining and the art of gathering, preparing, cooking and presenting food, not to mention conversation, drinking and the social side—animals must think we are mad.

Everything eats to live and without nutrition, animals and plants would die, but how sea creatures have evolved to get that meal that is quite amazing.

FANTASTIC FEEDING FACTS

Using the web-like skin between its arms, an octopus can carry up to a dozen crabs back to its den— all are doomed.

A baby grey whale drinks enough milk to fill more than 2000 bottles a day.

Dolphins and whales, like bats, use echolocation while hunting for food.

Whale sharks have the largest number of teeth in the world. They have more than 4500 teeth, with each tooth measuring about 3-4 mm.

The dominant beluga whale is called a mermistress; she looks after the young ones, often eating those that constantly misbehave.

By swallowing water, the pufferfish becomes too big for other fish to swallow.

Manatees may look fat and insulated, but the large body of the manatee is mostly made up of their stomach and intestines, cold water will kill them.

Dolphins have two stomachs; the first one stores their food and the second one is where digestion takes place.

It is possible for a mature dolphin to eat up to 13.6 kg (30 lb) of fish daily.

Dolphins often use a hunting tactic of circling the fish in a school so that they make a tight ball. Then they will take turns swimming through the centre, taking fish as they do. Soon, all that is left are sinking fish scales.

Walrus whiskers are called vibrissae and they have 450 of them. These whiskers are very sensitive and are used to help them find buried food.

The Bobbit worm is so named after the incident in the early '90s where Lorena Bobbitt cut off her husband's manhood with a kitchen knife. The hunting method of this giant polychaete worm has remarkable similarities to Mrs Bobbitt's method of attack. The worm rises out of the sand and remains motionless.

Although almost blind, through detecting movement it senses a fish and slams shut a pair of scissor-like mandibles. These worms can grow up to 30 cm (1 ft) long!

Digestion is the breaking down of large food items into their basic molecular makeup. If a shark takes a bite out of a surfer's leg— a big lump of flesh— there's all sorts here. Muscle, blood vessels, connective tissue, bone and bone marrow, skin, fat—all made up of biochemical molecules. Proteins, carbohydrates, lipids—all these are chemicals the shark needs to live. The process of digestion involves breaking down that lump of flesh into its smallest components so each individual molecule can enter the blood stream. These molecules then move into the sharks' cells to allow them to function and so allow the shark itself to function.

However, not all animals living in our oceanic wonderland have a need for digestion.

Most of us think of the humble barnacle on a rock, filtering small particles of food from the surrounding water. The beast, *Anelasma squalicola* is one bad barnacle; it's parasitic and eats sharks by fastening itself to their skin and draining nutrients from their blood.

So, the surfer's leg ends up feeding a barnacle – the food chain!

There are more than 50 species of sea lice, all feeding on the skin and blood of fish. One such louse, the tongue-eating louse, devours the tongue of its host. This beast crawls into its host via the gills to nibble away at the poor victim's tongue until—you guessed it—it falls off. It then attaches itself and pretends to be the tongue, taking nutrients from the fish.

Talking about tongues, a blue whale's tongue is so large that fifty people could stand on it.

The longest parasite in the world is likely to be a tapeworm in sperm whales, growing to over 28 m (90 ft) long.

FEEDING ON THE SPOT

If an animal is stuck on a rock, or a hard surface, the one thing it will not do its whole life is move to another location. It needs to grow and reproduce to pass its genes onto the next generation of its species. To do this, it must eat without moving around. Welcome to the interesting reality of the sessile filter feeders.

Two main strategies exist here: those that suck water into their bodies and extract any type of food present in that water versus those that capture plankton and other floating food via some sort of net, then pull it in to feed on.

One adult mussel can filter up to 70 litres (15 gal) of water per day. So, mussel farms with rafts measuring 12 m² (40 sq ft) can filter around five million litres (1.1 million gallons) of water per hour, per raft.

So effective are the mussel's filtration capabilities that there is ongoing research on using these shellfish to clean up polluted areas of water.

What exactly do these filter feeders consume? Anything that is in the water. Not only is there a vast range of plankton, but there are all sorts of other bits: bits of algae, broken bits of flesh from others feeding, slime from animals— the list is endless. If it's organic, its food.

Two problems facing filter feeders are firstly, that not everything in the water can be digested and secondly, that filtered water needs to be expelled far enough so it's not taken in again.

Here is a random question – when do unborn babies eat each other?

Sand sharks start with up to 12 embryos in the womb by different dads. When the mother gives birth, only one or two remain, at 1 m long. The babies eat each other in the womb. This allows only 1 dad to pass on his genes and results in a larger baby at birth, which is more likely to survive.

Back to our filter feeders.

The water is not just filtered by the animals, but also sorted by size. Particles that are too large are instantly expelled and anything within the correct size range is absorbed. Indigestible items are simply excreted.

The second problem is treated in two ways; the exit point for filtered water is located at the other end of the animal and simply flows back into the sea. Or in the case of the sponge, the exit tube is much smaller in width than the intake, this caused the outflowing water to be forced away from the animal, like putting your finger over the end of a hose pipe.

FANTASTIC FILTER FEEDING FACTS

Sponges have been filter feeding for over 700 million years.

Buried worms and shellfish will produce a sticky net, which acts like a sail in the water— plankton and other bits stick to it. The animal senses the drag on the net and sucks it in to feed.

Through evolution, barnacles have lost their arms and gained feathers, which they beat to push the water and food toward their mouths.

520 million years ago, a predatory shrimp traded in its claws to instead evolve arms resembling mobile nets for capturing food. This was the earliest-known large swimming filter feeder.

Such adaptations have taken place many times in earth's history, but only when species richness is large enough.

Changing a feeding strategy is a way to avoid competition for food resources. That sparked the rise of mobile filter feeders.

There are many strategies; sticky nets, feather-like arms, swimming with open mouths and gulping water— all to filter away and feast on the ocean's delights.

So, when is a filter feeder another filter feeder's food? It's all a matter of size. If you're small, then all you can filter out of the water are microscopic bits; if you're huge then you can filter out large bits from the water you take in.

The krill, a relatively large shrimp, filter feeding on small particles is eaten by a huge whale, filtering the shrimp out of the water it gulps. It's all a matter of size.

FILTER FEEDING GIANTS

There are 12 species of whales that filter feed.

Baleen plates line the whale's mouth and as water passes through them, the food sticks to these plates. Then, the tongue will wipe them clean and swallow.

Grey whales are bottom feeders, sucking in sediment and filtering out the juicy mud-dwelling animals.

Right whales are skimmers, they swim with their mouths open, skimming the surface of the water and filtering.

Lunge feeding is when a solitary whale swims under a shoal of krill or fish, then it lunges upwards, engulfing everything in its way.

Bubble net feeding is when whales, acting in a group, release air bubbles in a circular shape. This causes an underwater wall of bubbles that prevents anything inside the circle from escaping. The whales then take turns to lunge at the doomed shoal of fish or krill.

A humpback whale can contain 69 000 litres (15 000 gal) of water in one mouthful. That's 57 000 kg (125 000 lbs) of water!

When swarming, krill can attain densities of 30 000 animals per cubic metre of water.

The blue whale can consume 4 tonnes of krill or 40 million individual shrimps per day.

VICTIMS OF VENOM

One type of capture net employed is that of the jellyfish; instead of using mucus to trap its prey, it makes use of deadly stinging cells. Then the corpse is wrapped up and pulled in.

We have now entered the realm of the venom feeders. The oldest known animals containing venom are jellyfish, dating back to a known 560 million years (but it is thought they were around as far back as the first multicellular organisms at 700 million years).

The lion's mane jellyfish is recognised as the largest jelly in the world, reaching lengths of 2 m (6.5 ft) across. Fully grown adults can have 1200 tentacles up to 36 m (120 ft) long.

Cassiopea, or the upside-down jellyfish, has lost the ability to feed. It contains algae in its tentacles which, when growing, produce waste products and these in turn feed the jellyfish. Its swims upside-down to give the algae sunlight to grow and, thus, feed their host.

Take a jellyfish, mix it with a cassiopea and shrink it down to a very small size. Then we will have coral; it feeds exactly like the jellies by stinging zooplankton but is also full of algae to give it an extra boost.

A close relative of the jellyfish is the sea anemone, which uses an identical prey capture method to the jellies, only it hunts while remaining fixed in one place. With tentacles surrounding its mouth, anything caught is simply passed to the mouth for digestion; its waste is also passed out of the mouth. One way in, same way out.

STINGING SNACKS

So what eats corals, anemones, and jellyfish? They all sting. Well, a sting is only effective if it can pierce the skin. So, if you have bony plates as a mouth or very thick skin, jellies make a nice snack. Acid in the stomach will destroy any toxin before it can get to the blood stream.

Bite for bite, fish provide about 30 times more calories than jellyfish, but jellyfish exist in huge swarms and cannot swim away!

Ocean sunfish and leatherback turtles are known to be big feeders of jellies, often eating hundreds per day; well, they need to—they are really large animals.

Examining the DNA in stomachs of eel larvae, scientists found 76% of prey DNA belonged to jellyfish.

Looking at albatross droppings, scientists determined jellyfish made up 20% of their diet.

Placing mini cameras onto penguins it was found that 40% of their diet was in fact jellies.

Sea slugs eat anemones and corals. They ingest their stinging cells without triggering them. The cells pass through the slug's stomach, remaining intact, then move through the bloodstream to the outer skin, where they protect the slug! Anything touching the slug will be stung by a jellyfish sting—crazy!

THUGS FROM MARS

Thugs use brute force to get what they want. The thug of the sea has to be the mantis shrimp. With over 200 species, the large ones are dangerous to humans, so keep away! This creature's unique biology has given it the nickname, 'the shrimp from mars'.

Humans have three coloured light receptor cells in our eyes; our shrimps have 12. Each eye is on a stalk and can move independently.

Our mantis shrimp will choose one partner for life. With this one partner it will share its food, shelter and raise offspring.

Bigger species over 20 cm long feed on large crabs. They have 2 clubs as arms and can fire them out at 80 km/h (50 mi/h), shattering any shell they hit.

This movement happens with such rapid speed and friction that it is estimated, for a mere millisecond, the water around the club's surface will mirror the sun's heat.

The chemical make-up of the club has been analysed to find out why it does not break under such force. This led to the discovery of a new method of shock absorption, inspiring new techniques of carbon fibre manufacture for planes and other fast-moving machines.

INGENIOUS EATERS

A species of mobile snail called a whelk will use a tooth designed like a drill bit to slowly drill a hole into a mussel. To help the drilling process it excretes sulphuric acid to weaken the mussel's shell. Once it has broken through, the whelk injects enzymes to digest the flesh into a soup. The soup is then sucked out.

A species of starfish will crawl over a cockle and, unable to pull its shell apart, it quickly glues its tube feet to the shell, creating a very small gap. This gap is just big enough for the starfish to push its stomach out of its mouth and into the shellfish, where it digests and absorbs the flesh.

HIDEOUS HAGFISH

When looking at feeding in the sea we have to award the gold medal for the most disgusting to a deep-sea animal called the Atlantic hagfish. Are you ready for this? Here are a few facts…

The hagfish resembles an eel. Its skeleton is made of cartilage instead of bone, like a shark's. It has only a partial skull, no jaws and no scales. This fish is nearly blind but has a good sense of smell and touch and grows to 1 m (39 in.) in length.

It is covered in a special slime that is thick and sticky because of the stiff fibres it contains. The hagfish can produce slime at will and can fill a 500 ml jug in seconds. If threatened, the slime will clog the gills of any fish.

It can tie itself into a knot and pass the knot down its body to wipe the old slime away, ready for a new coat.

Depending on the population it can change between being male or female when it needs to.

It is a predator with no jaws. Around its mouth are hooks which will attach to a fish's side, the tongue then gets to work. It is like a file, rasping off flesh; the eel-like fish then bores into the body of the prey, scraping the flesh as the victim is eaten from the inside out.

Talking about eel-like fish, oarfish are the longest bony fish in the world. They have snakelike bodies and can grow up to 17 m (55 ft) in length; they are thought to account for many sea serpent sightings.

DEEP SEA DINERS

There are many species of deep-sea animals; ones that live on the bottom and ones that swim. Feeding is a joy for the bottom dwellers as most are filter feeders or deposit feeders (they collect food on the sea floor or buried in the mud). The advantage here is that there's a constant supply of food falling from the upper ocean 24 hours a day. There's so much of it that it's called marine snow—it looks like snow falling on a mountain, only underwater.

However, if you swim it's a different matter. The first problem is that populations are small, so finding swimming food is difficult. The second problem is that it's constantly pitch black.

Smaller animals have evolved to form very large colonies of individuals. If you have a stinging cell, then it's not worth having if you do not bump into something to sting!

To increase the chance of hitting a prey item, these animals form a colony of joined-up individuals resembling long strands of string floating in the water; the longer the colony, the greater the chance of hitting prey. Once successful, the digested juices are shared between all the individual animals.

Larger animals have developed different strategies to find food. The most common is the ability to produce their own lights through a chemical process known as bioluminescence. A light, produced by a special organ known as a photophore, can turn on and off at will.

Now add this to another adaptation, an enlarged head and jaw. With lights around the animal's mouth and head, unsuspecting prey are attracted, thinking 'dinner!'— only to be eaten whole. One problem here is a predator flashing for a dinner might just attract a larger predator not flashing for dinner!

As previously mentioned, deposit feeding is a method employed by many species that eat organic material from the sediment —basically anything that settles on the floor. Also, the microscopic area between sand grains contains a whole population of microscopic animals and millions of bacteria, all food.

FANTASTIC FACTS – SEAFLOOR FEEDERS

Sea cucumbers are often known as the bulldozers of the sea, as most species live in the sediment, moving through it like bulldozers on land. As they move, they simply open their mouths and pass the sand through their bodies. All the organic material is removed by their guts and the rest expelled as clean sand.

Sea cucumbers engulf sediment through their mouths, so how do they get water into their bodies to breathe? They suck it in through their bottoms!

Sea cucumbers inhabiting the deep ocean make up over 90% of all life found in the sediment. Their feeding habits ensure that the sand is turned over regularly and remains fresh.

Crabs' claws are a giveaway as to what they eat; large predators have one crushing claw and one thin one for slicing flesh. Algae eaters have flat scooped claws for

snipping and catching the algae. Filter feeders have feather claws. Just look at the shape of their claws to find out.

TOOTHY TALES

Round urchins are often algal feeders, grazing algae off the rock surfaces. The urchin is round and solid, yet the rock surface is unevenly shaped, often with crevices. The urchin's mouth has 5 jaws—5 cutting blades which can move, poke out and fit into any crevice or undulating surface. Problem solved.

So, if you're a fish and want to eat an urchin, you have two problems; one is spike avoidance and the second is how to get to the juicy bits behind the hard shell.

The Atlantic wolffish is a vicious looking beast, but it's harmless to humans. Its big front teeth protrude outside the mouth making it look particularly nasty. Well, it is certainly nasty to urchins! Its front teeth grab the urchin and crack the test (shell); spines are no match for the wolffish's thick muscular jaws. Then, the urchin is crushed by the large molars in the back of the jaw.

Want to know what a fish eats? Look at its teeth.

The sperm whale has the largest teeth of any whale. Only the lower row of teeth are seen; the upper teeth never pierce the skin. They weigh over 1 kg (2.2 lbs) each.

The white shark has over 3000 teeth at any one time. When a tooth is blunt, it falls out and another super-sharp slicer replaces it.

The deep sea fangtooth is the name given to two species. The largest is only 16 cm but the other species is only known as a baby; how big it grows, no one knows. Its two front teeth are so big it has special sockets running past each side of the brain for them to sit into. Even so, the teeth are so big that these fish can never fully close their mouths. This is the largest tooth in relation to body size of any fish.

Fish feed on everything you can imagine. There are filter feeders, algal grazers, plankton feeders, scavengers of dead bodies, invertebrates, diggers of sand, feeders of whole smaller fish— the list is endless. Their teeth give it away.

The way they catch their food is just as diverse; they wait in ambush, hunt in packs, scrape rocks and pretend to be food or just swim along waiting for something to pass by. Again, the list is endless.

However, they all have the same problem, which is that their prey is roughly the same density as the water. This means that the prey will move in the same way as the water. Close your mouth and you push water away, along with your meal. Four main strategies have evolved to solve this problem. All are highly complex methods and most fish employ two or more to acquire a tasty morsel.

Suction feeding is when a fish literally sucks in large amounts of water, bringing the prey with it.

Ram feeding is when the fish simply swims with its mouth open, forcing water and food through the mouth.

Protrusion is when the fish's jaws can move out of their normal fixed position and extend into the water, closing around the prey.

Pivot feeding is when the fish's head is anchored to a neck pivot, which moves around and then sucks in prey.

The razorfish, which grows to 20 cm, is the fastest feeding animal on the planet. A relative of the seahorse, it hovers upside-down. Using pivot and suction methods, it can suck in a shrimp in under 5 milliseconds. Sorry McDonald's, that's what I call fast food.

Grazers rely on a constant (but low) supply of food, algae, or zooplankton. They continually process this food through a long gut, allowing digestion to take place. They have small stomachs but long intestines. It's like a conveyor belt that hardly stops.

Predators have bigger stomachs yet smaller intestines, as their food arrives in large amounts but within longer time intervals.

Giving a fish a hand in the sea can be the wrong thing to do. Fish are attracted to an easy meal in the form of the bread you are holding. Yet while the fish take a break from their natural behaviour, predators can move in and eat the eggs they should be guarding.

Natural bacteria in the fish's guts may become useless as they cannot help to digest the unnatural food that was given. Over time these fish suffer from malnutrition because even though they are taking in food, it is not being digested correctly. Try not to feed fish in the wild.

REPRODUCTION AND LIFESPAN

YOUNG, OLD & BROODY

The shortest-lived fish, the pygmy goby, will never see 60 days on earth. 27 of its days are spent as larvae. In human terms about half of its life is spent in childhood.

The average lifespan of a dolphin is 17 years. However, some have been observed to live about 50 years in the wild.

The marine fish with the longest life span is the rougheye rockfish, found offshore in the North Pacific and living for a record 205 years. Artic species are thought to be longer-lived but are yet to be found.

The *Arctica* clam is the longest-lived animal species; one specimen was found to be 507 years old. Older clams must still be alive today.

The ocean sunfish lays up to 5 million eggs at one time.

A female oyster may produce over 100 million young over her lifetime.

Reproduction and life spans are intrinsically linked. The longer the lifespan, the longer the animal takes to reach sexual maturity. This is a really important fact when we consider fishing practices.

If we catch a species that takes 2 years to reach sexual maturity, then that species will recover from overfishing within a relatively short timescale. If we damage stock that take a long time to reach their breeding age, then that stock will be in real trouble and take many, many years to recover—if ever.

In the 1800's we overfished the great whales. By 1931 it was so bad that the first treaty to stop the practice was introduced. By 1984 total bans were in place, but a terrible loophole exists to allow some hunting. The Southern Ocean was declared a whale sanctuary in 1998 (although often ignored). So, in 90 years the whales have made a great comeback. They have not fully recovered, yet thankfully, a huge step has been taken in the right direction. But it took 90 years!

When looking at lifespans one thing is apparent, if you live to an old age in the ocean, you have done very well indeed.

Which factors in the oceans dictate how long an individual will live? Size, metabolism (how fast your body works) and where you live are but a few.

Larger animals tend to live longer because they grow too big to be eaten; even at birth, they are big enough to avoid many predators.

Blue whale calves are up to 8 m (27 ft) long and weigh approximately 3 tonnes (6 600 lbs). Safe, you bet, but not from killer whales!

Most large animals giving birth to live young tend to only have one or two offspring. Compare this to the 1 million eggs a cod will lay, with only 2 making it— we can see how being big is a good thing.

Metabolism is how fast your body works, or how fast a heart will beat. The faster it is, the shorter your life span will be. A heart can only beat a certain number of times before it stops. A faster-beating heart will reach that number in a shorter time, hence the shorter lifespan.

A small fish will be able to dart into the coral at very high speeds due to its fast metabolism; a larger animal will take longer to reach top speed but can maintain it for longer.

In 2015, researchers with the NOAA Southwest Fisheries Science Centre revealed the opah, or moonfish, as the first fully warm-blooded fish. Hunting at depths of up to 400 m (1300 ft), its warm blood allows it to move faster to catch prey.

All marine invertebrates are cold blooded.

Where an animal lives will determine how long it will live. Most sea creatures are cold blooded, which means their bodies are at the same temperature as the water they swim in. Warm is fast, cold is slow.

A barnacle will reproduce once a year in temperate waters, yet in the tropics it will reproduce every 14 days.

Life in the tropics is life in the fast lane. Life in the artic regions and the global deep oceans moves very slowly due to the cold temperature of the water. Slow metabolic rates ensures a very long lifespan for those living there.

FANTASTIC FACTS – SLOW GROWERS

It can take a deep-sea clam up to 100 years to reach 1 cm in length. The clam is among the slowest-growing, yet longest-living species on the planet.

The oldest living species of urchin was the red sea urchin, which was dated at around 200 years old. Living in the tropics, this is a real oddity—that's the beauty of biology, there is always an exception to the rule.

Polar gigantism is when animals grow to a huge size compared to their tropical relatives. Amphipods (like woodlice) in the tropics measure a couple of centimetres while their polar relatives exceed 30 cm (11.8 in.) This is due to a combination of cold-driven low metabolic rates and the high oxygen content of the water (cold water holds oxygen better than warm water).

This was also the case during the time before the dinosaurs, when the atmosphere's oxygen content was 30% instead of the 21% it is today. This enabled giant insects to exist— centipedes over 2 m (6 ft) long and dragonflies with 70 cm (28 in.) wingspans. Then oxygen

87

levels decreased along with insect size until 150 million years ago when we had another oxygen spike, but the flying insects remained small— why? The evolution of birds.

How old can these polar animals get? Well, we don't know. There could be a 1000-year-old clam, but not many people would want to find out, as to do so, the clam would have to be killed— not the way to go!

Whether you are alive for 8 weeks, 80 years or more in the marine habitat, the one thing you will try to do above everything else is to mate successfully. This will pass on your genetic code to further your species and can also encourage the evolution of a new species. This process, called speciation, typically takes between 2 and 20 million years.

However, it is not as simple as that, because many creatures will be able to reproduce without finding a mate. This method will be less likely to further evolution as the identical genetic code is shared by both parent and offspring. Genetic mutation will still occur, only at a vastly reduced rate. It will, however, give a species the advantage of taking control of a single area of habitat because its offspring don't go through larval stages and are not dispersed by currents.

All animals that exhibit asexual reproduction also reproduce sexually. Here we encounter the bizarre world of sexual and asexual reproduction in the oceans.

FANTASTIC FACTS - CRAZY CLONES

Parthenogenesis is a form of asexual reproduction in which an unfertilized egg develops into a new individual. Zebra, hammerhead, and blacktop sharks are known to do this when finding a mate proves to be difficult.

Corals can reproduce by budding— one coral will simply split into two smaller animals. This can increase the size of the colony. Is it a good method? Look at the size of coral reefs!

An anemone will walk over a rock and leave a ball of cells from its foot behind; from this, a new animal will grow.

A starfish will shed an arm, and soon one has grown into five. Great if you are being eaten— leave a couple of arms behind and off you go again.

Brittlestars have been found at densities of over 2000 individuals per square meter— all genetically identical. This strategy has outcompeted all other species from the area.

Fishery workers were instructed to cut in half any starfish they found eating the clams and oysters that they were farming. This doubled the population of predators as each half grew into a whole one. (Oh dear!)

OCEANIC ANCIENTS

Where do we find the highest number of the oldest species on earth? In the oceans—but why?

80 million years old—frilled shark
287 million years old— barnacle
295 million years old— shrimp
360 million years old—coelacanth
360 million years old—lamprey
450 million years old—horseshoe crab
500 million years old—nautilus

This is just a little taster of prehistoric species that are still living yet have remained unchanged through all that time.

The biggest single factor that causes a species to evolve into another is climate change. When the climate changes, species have to adapt, or they die out. Normal conditions allow this process to happen really slowly, but sadly, under the current conditions we will see many species becoming extinct in the next 50 years.

Looking at our oceans, particularly the deep sea, we find that conditions have not changed there since the first oceans were formed, so the animals have not had any pressures impelling them to change. Many previously-thought-extinct species will still be down there, doing the right thing – hiding from us!

SEX IN THE SEA

Sexual reproduction in the sea as well as the bizarre larval lifespans and behaviour of some of the creatures is truly like something from outer space. The best Hollywood director could not have dreamed up such magnificent behaviours. They have tried. Watch Avatar (2009) where the majority of plants and animals in the alien forest could be found on a coral reef on earth. Still, they came nowhere near to what we shall look at— crazy, you bet!

Meiosis is a type of cell division that reduces the number of chromosomes in the parent cell by half and produces four gamete (egg or sperm) cells. During fertilization, the sperm and egg unite to form a single cell, the number of chromosomes is restored in the offspring and sexual reproduction has occurred.

Sexual reproduction in the marine environment can be split into two wide ranging groups, each with subgroupings. However, the fact remains that we have two types of animal, those that can move to find a mate and those that cannot move but need to do their business anyway.

FANTASTIC MATING FACTS

Many fish can change their gender during the course of their lives. Others, mostly deep-sea fish, have both male and female sexual organs, so if by chance they

meet, irrespective of which sex they meet— they can mate.

A bamboo shark can store live sperm in its body for up to 3.5 years.

The male deep-sea angler fish is 2000 times smaller than the female— not only that, it's born without a gut and cannot feed. It quickly finds a female to attach to, then fuses with it, so the female's blood feeds it. Next, the fish basically turns into a sperm-producing bag that when used up simply falls off and dies.

30 clownfish can live in one anemone at a time; the largest is the female being tended to by the males. When the female dies, the largest male changes sex and takes over her egg-laying job.

The sea hare, a type of large sea slug, will attract a mate; they join and the one in front acts as the female. Then, more join and the chain gets longer and longer, with opposite sexes attached to the top of one hare and the bottom of the next, it's like a very long mating rope.

In one deep-sea squid species, the male will hover above the female and hold on with its hooks or beak as it does not have suction cups. These create cuts, 5 cm deep, in the female into which the male pushes sacs of sperm which the female then absorbs.

Female seahorses deposit fertilised eggs into the male's pouch; here they mature until the male gives birth to live young. One species holds the record for the highest amount of live young given birth at any one time—over 2000.

Our mobile creatures all reproduce sexually, and they produce two different types of eggs, those that are protected and those that are not.

Unprotected eggs are produced in the millions; these tiny eggs are shed directly into the water. Male fish will swim with the egg-releasing females whilst discharging their sperm into the trail of eggs, where fertilisation takes place. There the unfertilised eggs, the left-over sperm and the newly- fertilised eggs all join the plankton to disperse with the currents.

Such is the timing of these events that large filter feeders partial to plankton, such as manta rays, converge on the area year after year to feast on the new planktonic arrivals.

The fertilised eggs hatch, then go through different larval stages, all growing as quickly as they can. It really is a race against time. 99.9% of these mini-athletes will lose the race and end up in the stomachs of other animals. 0.1% will reach adulthood and win that race.

As if that were not enough to contend with, the timing is also coordinated with the currents, ensuring that when the larval forms reach their juvenile stage (small adult forms) they have returned to their natural habitat. If we get an unusual, prolonged wind it can change the direction of the surface currents just enough to push the young out of their habitat, only to die. As a result of these events, a whole year's population can be lost.

FANTASTIC FACTS – BREEDING DANGEROUSLY

Lobsters are very aggressive and cannibalistic— a problem when it comes to reproduction. The female locates a potential mate and moves past him at a safe distance whilst peeing. Her urine contains special chemicals that will charm a male. After she has done this a few times, the male will calm down enough to be approached and mating occurs.

Grunions are small fish that hurl themselves onto the shores that stretch from Southern California to Baja, Mexico. The female fish thrusts her tail into the sand to dig a hole and lays her eggs, leaving her head sticking out. A male then throws himself at her, wrapping around her head and shedding sperm into the hole to fertilize her eggs. All of this happens out of the water where the fish cannot breathe.

The male argonaut squid measures just 2.5 cm (1 in.), while the female is a sizable 30 cm (11.8 in.). With this ample size difference, the male is potentially a nice meal. So how will the male reproduce with such a large and intimidating female? It swims up unnoticed,

detaches its own penis from its body and fires the arrow appendage at the female. This sinks into the female, depositing the sperm. Such is the assemblage of mating males that it looks like an ancient battle with a sky full of arrows. The male soon dies from its activities and the female swims off, her mantle full of protruding arrows.

Moving on to look at animals who protect their eggs, what stands out is that these eggs are much larger. For this reason they are produced in much smaller numbers.

Here, the female lays the eggs and either or both parents tend to them, defending them from predators and fanning them with water to remove slime and give the egg a constant supply of oxygen. Development is normally quite fast and the young hatch in full juvenile form, already in their natural environment.

This approach has many advantages but, when hatched, the young run the gauntlet of everything that wants a snack.

MOONLIT SPAWNING

So how do animals that cannot move reproduce sexually? Broadcast spawning, in which sperm and eggs are shed into the water is the norm, but other factors can come into play.

Sponges can shed their sperm into the water in such density that it looks like smoke from a wildfire. However, the eggs remain inside the females. The female filter feeders then absorb sperm in the same way that they usually suck in food, yet instead of eating the captured cells they transport these to their eggs.

This strategy works because of the relatively large size of the sponge species, but what about mini beasts?

Timing is the factor here— normally a full moon. At the time when the full moon is high in the sky, the tide often switches direction. During this phase, the currents have the least power. Add to this the phenomenon of night-time calm, and we have a water body that has slowed down significantly. This will give any sperm and eggs in the water more time to meet than if they were swept away by rough seas and strong currents. Of course, it doesn't always happen— sometimes there's a storm— but that's the ocean for you.

The most-studied form of this synchronised reproduction is exhibited by corals, as described by marine biologist Tom Schlesinger in 2015 after a night-time snorkel in the Red Sea.

'Within minutes, as if someone had flicked an unseen switch, thousands of corals had released their reproductive cells, which rose through the water in a kaleidoscopic blizzard. Small fish darted in to feast. Bigger fish arrived to eat the smaller fish. It was a complete celebration of life. If I wasn't floating in the sea, I would have fallen to my knees.'

Not just one, but many species of corals will release their sperm and eggs into the water at the same time, within minutes of each other, on the same night every year. They all sit there ready, waiting, with the full moon being the main cue. As it rises, the corals wait, and they wait, until the first one releases. This sends out a chemical signal, like the starting gun at the 100 m sprint; once it is sensed they all release— the race is on!

Climate change is rearing its ugly head again; it has been discovered that this once annual event is being disrupted by factors yet to be determined, with climate change thought to be a major player. Corals in certain areas are now not releasing simultaneously once a year but periodically at different times. This is resulting in a greatly-reduced concentration of reproductive cells in any location at any one time. The effect is a greatly-decreased fertilisation rate, thus, a greatly-decreased settling of mature larva to colonise and further the species. This could be going on more often than we think, and it could have devastating effects for years to come.

Not all sessile animals have evolved to shed into the sea to reproduce. Some fertilise internally without moving.

The barnacle cannot move but needs to reproduce. It has the largest penis in the world relative to its body size—15 times bigger than its body—and it can have many of them! This is then extended out of the shell, guided by chemicals released by a nearby female; once located, the penis dips into the female's shell and deposits sperm.

ADVENTURES OF AN EGG

So once the sex is all done and dusted, how do the babies or larvae develop into young adults? This is just as strange as any other tale from our underwater wonderland.

The general rule is: one egg + one sperm = one baby; well, here we return to our biological exceptions.

Jellyfish lay eggs that look like mini anemones under a rocky overhang. As an egg develops it starts to resemble a mini jelly. This baby sits at the top of the stalk and, when ready, is released into the currents. Amazingly, underneath it is another baby, but not just one—there can be over 15, depending on the species. These are all then released one after the other. One egg + one sperm = 15 babies!

Whilst on the subject of jellyfish, reproduction and lifespan, the immortal jellyfish is worth a mention. In its adult form, it's happy bobbing around. Then it gets stressed, for whatever reason, and something amazing happens. A chemical process begins and the jellyfish starts to change. It settles on the bottom and reverts back to the polyp it started as— back to the fertilised egg stage. It then morphs back into its juvenile form, and once released, it grows again into adulthood, only to repeat the process when stressed. It is immortal; however, only until it is eaten.

A crab will transform through up to 9 different body shapes whilst developing as plankton, until it moults for a final time, and then it looks like a baby crab. This shape is unable to float and the juvenile sinks to the bottom to begin it adult life.

The larval forms of many marine invertebrates look nothing like their adult forms. The fertilised eggs float until they hatch, and an unrecognisable creature emerges. It has 3 functions to perform: 1) capture prey, 2) avoid being eaten and 3) stay afloat.

By the time the hatching takes place, the egg will have drifted far from the adult habitat required by the baby. If it settles here, it has 99.9% chance of dying.

Our baby crab could have very long spikes, it might have sail-like structures or anything else that will aid buoyancy, like gas-filled sacs. This also makes it look bigger that it really is (along with the spike effect we have predator avoidance). As it grows, it will shed its skin and an even weirder shape will be evident; more elaborate with larger sails or spikes—why? Well, the baby has grown, put on weight and needs a greater buoyancy system.

Then just as the time is right, the currents return to the adult habitat, where there is an abundance of food waiting. The final planktonic moult occurs, and our juvenile settles out and begins its benthic life.

This tale isn't just about crabs but any beast that sheds its reproductive cells into the open ocean. Crabs, jellyfish, urchins, fish— in fact any animal with this reproductive strategy.

So, what happens to the 0.1% that miss the adult habitat? 99.9% of the 0.1% will die or be eaten very quickly, but the lucky 0.1% of this population will get a foothold. They will grow and, in biological terms, the species will increase its distribution— biological success of the random kind.

Settling into the adult habitat is not a haphazard as it looks; for some species it is very precise and may take a few attempts.

For many species that do not move during adulthood, like a sponge, settling is a complex process. The baby starts to fall to the bottom, looking for chemicals released by the adults to help it recognise its correct habitat. No chemicals, and it will attempt to rise again, then fall in another area, looking for that precise chemical smell. It is a race against time as, at this point, the baby cannot feed. Once the right smell is found, it will settle, stick to the ground, and then shapeshift one final time into its adult form. It can then start to feed and grow. If the baby cannot find the correct cue, it will most probably die.

Again, of the very fortunate few who miss that chemical trigger and have to settle, some might just survive, once again increasing the natural distribution of their species.

And just like in the film Jurassic Park—'Life will find a way'.

HOUSING – MARINE REAL ESTATE

FANTASTIC FACTS – CRITICAL CORAL

The Great Barrier Reef, 2300 km (1430 mi) long, is the largest living structure on earth and can be seen from space. It is not one long solid structure but instead incorporates nearly 3000 individual reefs and 1000 islands. Its reefs are made up of 400 species of coral, supporting well over 2000 different fish, 4000 mollusc species and countless other invertebrates. It is the biggest city under the ocean (or so we thought…)

Although shallow water coral reefs comprise less than 0.5 % of the ocean floor, it is estimated that more than 90% of marine species are directly or indirectly dependent on them.

There are about 4000 shallow water coral reef fish species worldwide, accounting for approximately a quarter of all marine fish species discovered to date.

The chemistry of a coral skeleton is so like human bone that it has been used to help bone grafts.

HOMES IN HIGH DEMAND

When can you eat your house? If a dead whale sinks to the bottom of the deep sea. Here it will provide, not just a home for countless animals, but also a food source for them, often taking over three years to fully decay.

However, this is the beginning of the end for the colonising animals that made it a home, as many will die once the body has fully decayed.

The marine property market is very much like the housing market on land— strange, but true. In cities and towns (areas of high density) we find higher prices per square foot of land— why? Because of the competition between individuals for a house. As we move further out from these areas, we find the demand reduces and so does the cost. Taking it to the extreme, we find areas of land that are worthless, such as deserts and marshlands, with very few people wanting to settle there.

Coastlines and reefs are our cities and towns; then we move outwards to the shallow seas, where plenty of life exists but at much lower densities with fewer species. Then to the open seas, which are our deserts— vast spaces of lifeless water. A rough analogy I admit, but it paints a picture of real estate demand in the sea.

The Australian barnacle invaded northern European shores after it was brought to the area by ships. It has a thin edge to its shell, and as it grows it will push this under the shell of the European barnacle, popping it off the rock. More land for the winner.

Losing your home in the marine environment is most likely going to cost you your life.

There are animals that will build their own houses by sticking things together or secreting chemicals to form a structure. Others will dig and dig, bore out rocks or settle on them, use other animals as their homes (both their exteriors and interiors) and will seek out any man-made hard surface— basically if it's in the sea and it can be used, it will be. In a very broad sense, we can divide these into two different areas: man-made and natural.

ON BOARD A SHIPWRECK

It takes about 7 years from the sinking of a ship in shallow tropical waters for it to become a coral-encrusted reef-like structure.

It's a wildlife bonanza when a ship sinks, but it is also a very complicated colonisation process. As to who eventually claims a section of the new land, it's like a new development of luxury apartments— demand is high. So what are the factors at play that allow some to succeed where others will fail?

Let's look at the physical factors that eventually determine who will live in a house like this. The first is the movement of water over the wreck. The current may flow faster on one side of the wreck than on the other; it will slow down inside the wreck and along its different structures. In fact, the current's speed will vary in many locations. In some areas the water may flow smoothly, but in others it might mix like a washing machine. Add to this the light; many areas will be in full sunlight, others in partial or full shade and others in total blackness.

So we have one habitat, and within that, we will have countless mini-habitats all excluding and including different species.

Then we have the different species that will settle and fight for every available section of space, as well as the fish, which will move in to graze and feed on these settlers. Furthermore, there are larger predatory fish, circling the outer edges. Add to this the many mobile scavengers that will crawl over the surface and you have a bustling metropolis of life where very little existed a few years ago.

The first to colonise are the algae, which in turn provide food for the grazers; in come the fish, snails and urchins. Where the algae cannot grow, we find the dark- loving animals like sponges, worms and others.

After a year, the first soft corals appear and different encrusting invertebrates start to flourish. After 3 years we start to see hard corals— the reef builders that are slower -growing but pack a sting that will kill any other animal trying to overgrow them. After 7 years we have a reef-like structure on the surface, but in the dark, we also have countless species thriving. 7 years to build a city— not bad.

FANTASTIC FACTS – UNUSUAL ABODES

Up to 80% of all erosion in a reef system is caused by urchins. They chew away at the rock, forming small

tunnels. As algae grows over the surface, the urchins eat it— safe in their bored-out houses.

If you walk along a pebbled shoreline, you will find many pebbles pitted with small holes. These were once the homes of a small clam-like bivalve called the piddock. It bores into a pebble to remain safe, then filter feeds, never leaving its sanctuary.

Some fan worms secrete a hard calcareous material and construct tough tubes in which to live. Certain small species can be so densely populated that they form large worm reefs. Even then, they are not safe, as rock-boring sponges can infest the reefs and eat through their tubes, causing partial or total collapse of the reef.

The Ross Ice Shelf in Antarctica is the world's largest floating body of ice and can be over 330 m (1000 ft) deep. Drilling through it revealed strange undersea creatures, including upside-down fish. Europa, one of Jupiter's moons is thought to be the same—water under ice. Are there underwater creatures there too?

HYDROTHERMAL MARVELS

Tube worms come in all shapes and sizes, with the majority having soft tubes. In 1977 a new species of giant tube worm was discovered around a hydrothermal vent. Not only was this a new worm species, but it was, in fact, the first time a hydrothermal vent had been seen by humans. This was the most important biological discovery of the 20th century, as it was not just a species that was discovered, but a whole range of new ecosystems— the hydrothermal vents of the deep ocean.

Looking at the real estate that is the hydrothermal vent in human terms, you would not be wise to buy one; a vent only lasts about 30 years before it turns off and dies. When it does, all life around it dies too. Just like living in a house on the edge of a soft rock cliff—you know one day it will fall, but you live there anyway.

On the edge of the property, we find hot water shimmering through the floor, bringing dissolved nutrients with it and creating a localised rise in water temperature.

At the centre we have our skyscraper, a chimney constructed from sulphides and other chemicals that erupted from the earth's crust. In the penthouse on top of the skyscraper, we have the vent. Here super-heated water erupts at 350^0C, bringing with it a cocktail of volcanic minerals that cool and build our structure.

FANTASTIC FACTS – VOLCANIC VENTS

99.9% of the planet's ecosystems are based on a food web with the sun providing the initial energy. This is the foundation of life on earth.

Life around a hydrothermal vent is based on sulphide compounds spewing from the vents. Bacteria feed on these and, like animals grazing on grass, the vent animals graze on the bacterial mats.

Many species do not feed at all; instead, they house sulphur-feeding bacteria within their tissues. The bacterial waste products in turn feed the host animal.

Since 1977, 590 new vent species and nearly 300 giant worms have been discovered, but fewer than 50 vent sites have been studied in detail. There are estimated to be one vent for every 2–20 km (1.2–2 mi) of ocean floor at any one time.

Vents exist in the volcanic areas of the oceanic crust where tectonic plates split apart. When the plates move, they cut off the water supply and the vent dies, but with that movement another new one starts.

The Von Damn vent field is truly unique; it is a series of hydrothermal vents made from talc (magnesium silicate—the same chemical as baby powder), towering as high as 75 m (250 ft) from the sea floor.

It is thought that hydrothermal vents are where life on earth really began, rather than in the shallow seas. Because of the ever-shifting nature of the vents, we will never know.

MARINE MOBILE HOMES

Flotsam and jetsam—what's the difference? Flotsam are items or goods found floating on water that were not deliberately thrown in but may have come from an accident or shipwreck. Jetsam are items that were intentionally cast into the sea.

Humans like to roam around our ecosystems and go on holidays in recreational vehicles of all shapes and sizes, towing caravans and touring vast areas as we explore exciting new lands.

What is our marine equivalent? A plank of wood, a pallet, a barrel— really anything that is quite large and floats. It will be constantly on the move, travelling to new areas where its inhabitants might not venture under normal circumstances.

The most common settlers on the object's underside are large goose barnacles, and between them we will find sponges and other small sessile beasts. Around these we will find little fish, which in turn attract large fish.

Eventually the object will drift to new lands that these species will colonise, or it will float to warmer or colder places, where the fish swim away. For the attached creatures, it is the end of their travels.

Such is the attraction of drifting objects that fishing companies have intentionally set out large rafts, equipped with location beacons so they can be found in future. The larger the object, the more fish it will attract. Put quite a few of these together and you have a very large fish magnet. After some time drifting, the fishermen move in and simply collect the shoals of tuna attracted to the area.

FANTASTIC FACTS – REMIPEDE SURPRISE

Remipedes were thought extinct until a discovery in 1979. In 1981 they were fully-identified as living species; 19 species are now known.

They have a special housing need, requiring an area above an aquifer (permeable rock above a body of very salty ground water). This water seeps upwards into the sea, creating an area of localised high salt concentration.

Remipedes are the only venomous crustaceans with 2 elongated fangs to sink into prey, injecting toxins and digestive enzymes— it's a good job they are only 4 cm long!

DEEP SEA CITIES

The oldest human building dates to 4000 BC, so it is around 6000 years old. The oldest marine habitat created by animals is a deep-sea coral reef believed to be over 40 000 years old. However, that could be a very young estimate.

Just like hydrothermal vents, these are a new and exciting discovery, only having been known to exist for the last 30 years. As technology develops, we are learning more every day.

These reefs are built in exactly the same way as shallow water reefs, but here the corals do not rely on sunlight (good job, as no light reaches these depths), instead they depend on capturing plankton from the water.

Deep-sea reefs occur at depths of between 2000 and 6000 m (6500–20 000 ft). They are found in waters all around the globe— at that depth, the temperature and conditions are much the same everywhere.

There are around 6500 animal species associated with shallow water reefs and they have been well studied. In the past 30 or so years there have been over 3300 species of deep-water corals identified; the true count is not known and that's just the corals! Add to this the enormous number of fish and invertebrate species that will be found there. We are in the process of investigating the most diverse cities in the globe.

The corals form colonies in which new polyps grow over dead polyps. The oldest colony that has been analysed is 4265 years old, making it the oldest living structure in the marine environment.

These reefs often stretch for over 40 km (25 mi) and, as such, create whole cities for life to colonise. Not only that, they extend up to 35 m (115 ft) into the water column.

Imagine a futuristic city with huge towering buildings and hover-cars buzzing in and out between them; that's exactly what evolution has created in the depths. Tall corals stretching upwards, fish shoals darting in and out, crabs, shrimps, lobsters —the list is endless— all scouring over the surface. Truly an amazing, dark place to live.

Sadly, cities on land with ancient and wondrous architecture can be ravaged to partial destruction by war or other attacks. Similarly, these ancient marine wonders are in just as much danger when a trawler net scrapes through the reef, destroying everything in its way. 40 000 years of undisturbed growth taken in 10 minutes. More protection is needed for these delicate animals.

KELP AND CLIMATE

Where do we find equivalent tall structures in the shallow seas? It's another futuristic city, but one we can swim through along with the locals.

Giant kelp is the fastest growing organism on earth, growing at a rate of 60 cm (2 ft) per day to a total height of 45 m (150 ft).

The Tasmanian kelp forest has only 5% of its original area remaining in the wild. Such is the devastation, that it is Australia's first federally-listed endangered marine community type.

It has been predicted that ocean currents will shift with climate change. The East Australian current, bringing the nutrient-rich cold water that kelp needs to thrive, has indeed moved and been replaced by warmer water. Along with the invasive long-spined sea urchin (a kelp eater), these are the major players devastating Tasmanian kelp forests. Climate change a myth Mr Trump?— please look at the evidence!

MARVELLOUS MANGROVES

In cities, people like to reside in basement apartments, or those just under ground level. This has many advantages; it's easy to live there (only a couple of steps), and in hot climates it's the coolest place to be. Just like humans in cities, many tropical marine species like to live in the basement— well, in the roots of the mangrove tree.

There are around 110 mangrove species, ranging in height from 2-10 m (6.5- 33 ft).

Mangroves are found on the saltwater coasts of 118 tropical and subtropical countries, totalling more than 140 000 km² (87 000 sq mi).

Two thirds of the tropical fish species that are commercially fished spend part of their lives in mangroves.

Blue carbon is the term for carbon taken out of the atmosphere by plants in marine and coastal

ecosystems. They can be up to 10 times more efficient than land ecosystems at absorbing and storing carbon. This fact makes them critical in the fight against climate change.

Mangroves are an important blue carbon ecosystem, yet their unique wood makes them a valuable resource. Thus, they are being cut down on a global basis.

Who lives in the basement? Well, who does not? It's a really great, relatively safe place to be. The intertwining roots are impossible for large predators to penetrate, so it's a nursery area for countless species of reef fish. Predation rates are much lower than if they were growing up on the reef. Even baby sharks take refuge.

The invertebrate life is also astounding; everything you can think of burrows into the mud, takes up residence between the roots and attaches itself to them.

One snail even knows the time of the tides and will start to climb the root just before the tide comes in, escaping the water. Although the time of the high tide changes every day, this snail has a built-in clock to predict them.

Mangroves protect the coast and reefs. Their roots form a barrier— a buffer zone to stop the waves removing the shoreline. Those that were destroyed had all their mud washed away; it went out to sea and smothered the coral reefs—a double whammy.

OUT OF THE SHELL

95% of the human population will move house at least once in their lifetime. It can be an exciting time, but it is always stressful. For one species, moving house is a regular occurrence, but if it fails, it won't move house again!

Hermit crabs live in empty snail shells. There are over 800 species, of which 15 live on land.

If a hermit crab grows too big for its shell, it cannot retreat into it far enough to avoid a predator's beak.

If shells are in short supply, then the market for these homes becomes very competitive; so much so, that the occupants will often fight to the death for a house.

When we move house, we are often in a chain; one person moves in and the other resident moves out and into their new house, while that outgoing resident moves out to another, and so a chain is formed.

Hermit crabs will form a vacancy chain where up to 20 individuals all line up. Here, the largest will move house, leaving an empty shell for the crab behind it. This hermit then moves, leaving a smaller shell behind and it repeats—get the picture?

SUPER-EFFICIENT SEAGRASS

Town planners, don't you just love 'em! They have a difficult job, but like every job, there are good ones and not-so-good ones. Across the UK in the late 1960's and 1970's, town planners got to work and created whole areas with exactly the same designs of house or flat.

So where in nature do we find houses that are exactly the same, covering a huge expanse of sand? In the seagrass meadow.

Seagrass is the only flowering marine aquatic plant; it has a root system to obtain nutrients. Algae do not have roots, instead they have holdfasts to anchor themselves and must take in nutrients directly from the water, through their fronds.

Seagrass appeared 100 million years ago, and today we know of 72 different seagrass species. The grass forms dense underwater meadows, which are large enough to be seen from space.

In most areas, seagrass beds only contain 1 or 2 species, but in the Indo-Pacific we can find 14 species together. Only the polar seas are devoid of seagrass.

Seagrasses are referred to as the "lungs of the sea" it has been measured that 1 km^2 of bed will generate 10 litres of oxygen per day.

This plant is highly important in the war against climate change, as it is highly efficient at removing

carbon dioxide. Thus, it reduces ocean acidification. Yet, humans are destroying it rapidly.

The world's seagrass meadows can capture up to 83 million tonnes of carbon each year.

While seagrasses occupy only 0.1% of the total ocean floor, they are estimated to be responsible for up to 11% of the organic carbon buried in the ocean, effectively cleaning the atmosphere.

4000 m² (1 acre) of seagrass can remove 335 kg (740 lbs) of carbon per year. This is roughly the same amount produced by a car traveling 6400 km (4000 mi).

LIVING HOMES

Such things as living homes also exist in our marine world. By now, we all know about the importance of space and a hard surface on which to live. Well, the skin of animals or the fronds (leaves) of algae all provide surfaces on which to live—living real estate.

But is this just real estate or relationships between different species of animals co-existing together?

RELATIONSHIPS BETWEEN SPECIES

Two different species that live together and depend on each other throughout their lives have a symbiotic relationship.

The anemone and clownfish (*Finding Nemo*) have a very famous marine symbiosis. The fish protects the anemone from predators and receives a safe home in exchange. Clownfish mucus resembles that of its host, so it is not recognised. Happy days!

There are three types of symbiotic relationship that have been identified. Let's look at them in more depth.

PESKY PARASITES

Rhizocephalans are bad barnacles; they castrate the crab they have settled on, then trick the crab into thinking it is carrying eggs (the barnacles). The crab will protect its 'eggs' all its life!

This is parasitism, which involves one organism gaining from a relationship to the detriment of its host. It is commonly thought that parasitism leads to the host's death, and indeed, sometimes it does. However, this is not usually the case. The host's death will give no advantage to the parasite. Either the parasite will die too, or it will be released to float around looking for another host to infect— a risky strategy.

The pearlfish loves the sea cucumber. This tiny fish swims into the cucumber's bum to hide from predators. Whilst in there, to thank its host, the fish slowly eats away at its internal organs.

We often think of parasites as very bad organisms. Seen from a biological viewpoint with no human emotions, they are a very important part of any ecosystem.

Often, parasites will infect a host that is high up in the food chain or indeed the apex predator. Here, some species will cause the demise of individuals. This action stops any one species from becoming too dominant. Such is their importance that in some areas the whole biodiversity and richness of a habitat is calculated from the number of different parasites present at any one time.

FANTASTIC PARASITE FACTS

Here's one for you: there are at least 300 parasitic worm species waiting to infect you, and there are many more, in fact thousands, of different human parasites. Have you got any? You bet!

Parasites that live inside their hosts are called endoparasites and those that live outside their hosts are called ectoparasites.

There are more than 550 species of sea lice found around the world, feeding on fish.

Sea lice are considered pests in aquaculture, where they cause skin lesions and render the host vulnerable to other diseases.

Do you like eating raw fish—sushi? Well, please check for worms in the flesh! One study published in 2020 found a 283% increase in worm infestations since the 1970's on a worldwide basis. Once eaten, the worm attaches itself to our guts, causing vomiting and toilet runs for up to 3 days.

KINDLY COMMENSALISTS

One large jellyfish observed over a reef was found to have a whole school of baby fish (shads) living under its bell, far from the stinging tentacles. Also crawling over the jelly were a species of brittlestar and 2 species of mini-crabs. The jelly conferred protection to all but gained no advantage for itself.

This is commensalism: a symbiotic relationship between two or more different species where one or more of the species benefits and the other (normally larger) species is neither helped nor harmed.

A very common, often-observed example of this relationship is the great number of barnacles residing on the sides of whales or on turtle's shells.

When the imperial shrimp finds a sea cucumber, it will often use it as public transport, holding on until the cucumber stops moving. The shrimp will then disembark to hunt and climb back aboard to travel to the next feeding ground. Tickets please!

The remora, a fish with a large suction cup on its head, will stick to a shark's side and grab a lift to wherever the shark is going.

THE FEELING IS MUTUAL

Mutualism is when both partners benefit from a relationship; in the extreme, if one partner dies then the other will not be able to survive.

The pistol shrimp, which is nearly blind, digs and maintains a burrow. It shares the burrow with a goby. Whilst digging and excavating sand, the shrimp is vulnerable to predators. Luckily, the goby maintains watch. If a predator lurks, a flick of the goby's tail against its antennae tells the shrimp to retreat to its burrow along with the goby, both to hide. The shrimp also gains food while the goby gets somewhere safe to lay its eggs.

Cleaning stations are common on a shallow reef and are manned by an army of cleaner shrimp. Fish will line up in queues to wait their turn, just like humans in their cars at a carwash. The fish present themselves to the shrimp, which crawl all over their bodies and into their mouths and gills, picking off parasites and cleaning. The shrimp feed and the fish are cleaned. Often the fish clients are large predators who would not think twice about eating shrimp, but these cleaners are definitely not on the menu.

The decorator crab nips off pieces of sponge and attaches them to its body as camouflage. The sponge grows and benefits by being mobile and taken to many food sources.

The boxer crab finds a distinct species of anemone, which it tickles, causing the creature to move onto the top of its claw. Here the anemone feeds on the small, fleshy leftovers of the crab's meals. If a predator comes knocking, the crab lifts its anemone–tipped claw with a boxing action to warn it off.

There are hermit crabs that wear anemones on their shells; when they move house, they move the anemones too.

Pilot fish swim where others dare not— at the open mouths of sharks. There they pick at flesh that is caught between the teeth. The shark gets clean teeth and the fish a meal.

Some relationships are so mutually dependent that to lose your partner is to lose your life.

Reef-building shallow water corals have huge amounts of algae living inside their tissues. The corals provide nutrients (their natural waste products) and protection. The algae provide the corals with food in the form of algal waste products and dead algal cells.

If corals become stressed due to high temperature (one degree can be enough) or a very small shift in the pH of the water, this triggers a deadly response. No one knows why, but the corals expel all the algae from their bodies. This causes the corals to die, leaving behind pure white skeletons; it is known as the bleaching effect.
Again, with global warming we are seeing more and more bleaching. The reefs can repair, but if large species of algae colonise the dead coral skeletons, the repair is lost, and soon other species that depend on the living reef vanish too.

CRUCIAL KEYSTONES

Here we move to the phenomenon of the keystone species. Think of a brick or stone arch: at the centre of the upper arch is one stone; remove that stone and the whole structure collapses. Now think of a whole ecosystem, containing a wide variety of species in natural balance. If we remove a certain species, it has the effect of collapsing and changing the whole ecosystem. This is the keystone species.

The common starfish is a mussel eater; its predation allows other animals to take hold of the rock surface that the mussels like. One study removed the starfish and within a year, over 50% of the rock-attaching species were lost, due to mussels taking over. The starfish were then reintroduced and within 18 months the natural biodiversity had returned.

Often cited as the most important keystone species in the oceans, it is also the most hunted— the shark. Sharks weed out the sick and diseased, they maintain balance and reduce fish infections (by eating). They are termed the global oceanic keystone species.

Sharks are the only animals that never get sick and are immune to every known disease including cancer.

The average adult human will eat 1000 kg (2200 lbs) of food per year. One species of parrot fish will eat 4450 kg (9900 lbs) of material in the same time span. Its teeth are made from the second hardest biological material—harder than silver. They have to be, because they crunch at corals all day. They remove the surface of the corals, which allows new regrowth and, importantly, takes any algae away too. Without this species, coral reefs would soon become algal reefs and lose their incredible biodiversity. Parrot fish also poo out clean sand and so help maintain the sand content of the area.

Remove krill and the entire Arctic and Antarctic ecosystems will collapse; guess what some companies are thinking of fishing for?

WHAT'S GOING TO KILL YOU?

As a species we are unique; the one thing that intrigues us more than anything else is what could cause our untimely demise. In other words, what can kill us?

What is the single animal in the ocean that instils the greatest fear, yet is the one least likely to cause death?

Let me present a quote from a film that caused widespread panic among humans and widespread hunting of large sharks: "…it's all psychological. You yell barracuda, everybody says, 'Huh? What?' You yell shark, we've got a panic on our hands on the Fourth of July." – Jaws, 1975.

The International Shark Attack File is published on an annual basis and collates every recorded attack from the international community.

The ISAF documents two types of attack:

Unprovoked attacks which are incidents in which an attack on a live human occurs in the shark's natural habitat with no human provocation of the shark.

Provoked attacks when a human initiates interaction with a shark in some way. These include instances when divers are bitten after harassing or trying to touch sharks, bites on spearfishes, bites on people attempting to feed sharks, bites occurring while unhooking or removing a shark from a fishing net or any interaction that can provoke a response.

The latest figures from the ISAF are listed below and contain data from unprovoked attacks only.

2010	82 Attacks	6 Fatal
2011	79 Attacks	13 Fatal
2012	83 Attacks	7 Fatal
2013	77 Attacks	10 Fatal
2014	73 Attacks	3 Fatal
2015	98 Attacks	6 Fatal
2016	81 Attacks	4 Fatal
2017	89 Attacks	5 Fatal
2018	68 Attacks	4 Fatal
2019	64 Attacks	2 Fatal
2020	60 Attacks	12 Fatal

Humans kill over 100 million sharks each year!

So in the time period above, sharks have killed a total of 72 people, yet humans have killed 1 100 000 000 sharks.

However, the fact remains that if you put the average person into the water with a face mask on, tell them to look outwards to where the sea blends into darkness, they will usually think of one particular big fish swimming towards them. It's not so much the fear of dying, but the most prehistoric and primal human fear of being eaten alive that comes into play.

It has been recognised that any shark longer than 1.8 m is a potential danger. Three species are repeat offenders: great whites, tigers and bull sharks, mainly because they hunt near the shore. The oceanic whitetip, a cruiser of the open sea, is also high on the list, but encounters are rare due to their isolated nature or not recorded due to shipwreck survivors simply not making it!

Luckily for the unlucky, 99% of sharks will only bite and let go —humans are not their normal prey. Unluckily, large sharks have large mouths; one bite is enough to cause enough damage to be fatal.

MONSTERS OF THE DEEP

The oldest recorded shark attack took place in Japan, near Osaka Bay in what is now called Seto Inland Sea. The skeleton is over 3000 years old, and analysis revealed that this was a very vicious attack with over 790 deep serrated wounds and parts of the skeleton completely missing!

The worst shark attack in history occurred with the sinking of the warship USS Indianapolis in 1945. After being torpedoed by a Japanese submarine, it is estimated that of the 1196 men aboard, 900 escaped into the sea alive. After 5 days, 317 survivors were rescued, and of the men that died, most succumbed through injury or lack of water; it is estimated that up to 150 were taken by sharks.

There have been no fatal attacks by killer whales in their natural environment. However, since the early 1970's, there have been numerous attacks— some indeed fatal— by killer whales in captivity.

The cause of these attacks is still being debated. Were they accidental or did the whales know exactly what they were doing? Killer whales should not be in captivity. The whole idea of orcas in captivity serves no purpose other than to entertain humans and generate cash. Please avoid these places.

From July to October 2020 there were at least 35 reports of orcas attacking boats off the Atlantic coast of Portugal and Spain. The attacks involved nudging, biting and ramming. A small group of orcas are believed to be responsible, however no humans were injured. It is thought that the orcas were playing— if they were hunting then the small boats would not have stood a chance!

Squid— do they or do they not? In the 1790's, a becalmed trading ship was going nowhere. The Captain, Jean-Magnus Dens, ordered men to sit on planks and clean the side of the ship. A giant sea monster surfaced and took two men down, an arm wrapped around a third, but the sailors cut it off. The third sailor died of his injuries later that night.

The Humboldt squid, a 6-footer, swims in huge shoals and rises to the surface at night in the waters all around Mexico. Aggressive animals would be capable of stripping the flesh off a human in less than one minute. There have been no certified attacks, but many fishermen have not returned from night fishing, fuelling tales of flesh-stripping squid.

One other animal that will bite and kill is worth a mention, the saltwater crocodile. Growing to over 6 m (20 ft) and living to over 100 years, these are very large, aggressive animals. Unlike sharks, the crocodiles will take their victims and consume every bit of them. This has been well documented, with whole-body remains being recovered from the attacking monster's gut. Attacks in inland areas are more common than attacks at sea. However, people have been taken whilst swimming and snorkelling, although the incidents are very, very infrequent indeed.

Collating the facts displayed above leads to one conclusion: large sea animals are highly unlikely to cause fatalities. Instead, it is the smaller animals, including single celled algae, which we must observe. The small have a great impact on humans entering the sea or indeed eating produce from the oceans. You would be extremely unlucky to be killed by a large predator and much more likely to meet your maker as a result of the smaller beasts, which attack through defence or by accident rather than through hunting.

KILLER CHEMICALS

Since 1890 it is reported that 17 people have been killed by stingrays. Steve Irwin, known as the Crocodile Hunter on the Discovery Channel, was pierced in the chest by a stingray's tail spine, causing heart failure.

There is no official record for deaths caused by jellyfish stings. It is estimated that 20-40 people die each year around the Philippines alone; the number of worldwide deaths is unknown.

This may surprise you, but in 2018 a fisherman was killed by a sea snake bite—the first in 80 years to cause a fatality in Australia. Sea snakes, despite the fear they might instil, are not the venomous killers their reputation implies. But don't handle one— you might just become the next victim.

Even venom cannot protect them—sea snakes are preyed upon by eels, sharks, large fish, sea eagles and crocodiles.

Sea snakes can breathe through their skin, absorbing up to 30% of their required oxygen. The left lung of a sea snake is enlarged, extending down the length of its body. This controls the animal's buoyancy and allows it to spend up to 8 hours underwater.

Now we reveal the world of venoms and toxins, chemicals injected or ingested into your body. These chemicals will bind to your nerves, causing one of two things: either the nerves will speed up and react uncontrollably, or the chemicals will block the nerves, stopping them from sending signals.

There are two organs in the body that will cause rapid death if they stop working—your heart and lungs (not counting your brain, of course). Of the two, your heart is the main target for our invading chemicals. Slow it down and it will stop; speed it up uncontrollably and the result is the same—a quick demise.

We have two different chemicals at play, the venom, and the toxin. Venom is a poisonous substance secreted by animals and typically injected into prey or aggressors by biting or stinging. Venoms often contain a blend of different toxins which enhance the overall effect. So, when you are stung you receive a cocktail of different toxins.

A toxin is a substance that is synthesised by a plant species, an animal or micro-organisms, which is harmful to another organism.

It's not just what enters your body that matters, but how toxic it is and also how much toxin is contained in one dose. Think of the sea snake—it is highly venomous but only delivers a very small amount in its bite— hence the low death rate. There's quite a lot to think about here.

Venoms or toxins can both be deadly; however, a venom is a huge molecule and, as such, can be neutralised in the body by anti-venoms. In contrast, as a toxin is very small, nothing can be done to stop it; only medication can relieve its effects.

FRIGHTENING FISH

The boxfish is a slow swimmer, but it secretes a powerful toxin from its skin when stressed. So when a large mouth surrounds the boxfish, out pours the toxin. Faster than the predator can swallow, the toxin lodges into its gills, stopping it from getting oxygen. The boxfish is spat out and the predator dies.

A pufferfish is harmless unless eaten; its guts contain bacteria that is collected from the algae it eats. These bacteria secrete a toxin—tetrodotoxin—that is deadly if ingested. Over 60% of people who get ill from it will die. Tetrodotoxin is up to 100 times as deadly as the venom of the black widow spider and over 1000 times more deadly than cyanide.

Fish are the obvious venom-packing sea creatures, but relatively few species will cause fatalities. Most will cause you intense pain— way into the 10 factor on a scale of 1 to 10— but you will live. However, the unlucky few who do not get that anti-venom are in real trouble.

Fish venoms are defence mechanisms that are not used to capture prey. So how do we get stung? Either by handling them, or most commonly, by treading on them.

Dead fish can sting; venom can be stored in sacs next to a spike. If the spike enters a body, the venom sac is ripped open, allowing the venom to enter the bloodstream. I know this because I was stung by a dead lionfish.

Hurt?

Well, I know what a factor 10 is!

There are 5 species of stonefish, ranging from the Red Sea and Indo-Pacific to Australian waters. They are the most venomous fish alive. No records have been kept for fatalities from stonefish, but they are known to occur.

DEADLY STINGERS

Here we will leave the large animals with backbones and look towards the relatively smaller invertebrates. Some possess extreme toxins to capture prey and others for defence, while certain highly dangerous toxins are not utilised by the animals at all—they only come into play when eaten, as with the previously mentioned pufferfish.

On 20 January 1955 in North Queensland, a 5-year-old boy was enjoying a swim. He was stung and soon succumbed to the venom. A scientist caught 3 species of jellyfish; one was an unidentified species and unusual because it was transparent and box shaped. Not only was a new species discovered, but also a new genus (grouping) of jellyfish.

Since then, many species have been discovered, ranging in size down to the thumbnail box. They inhabit all tropical oceans and, concerningly, are moving into temperate waters as the sea temperature changes.

All jellies have the same mechanism to deliver their venom— a harpoon. The nematocyst is a cell kept under pressure; it has a hair trigger that if pressed causes the cell to open and fire the harpoon into the flesh of its victim. Then venom is pumped into the body.

FANTASTIC JELLYFISH FACTS

Depending on species, a fully grown box jellyfish can measure up to 30 cm (12 in.) in diameter, and its tentacles can grow up to 3 m (10 ft) in length.

The thumbnail is the name given to the smallest box jellyfish—lethal despite its tiny size. There are about 12 tentacles on each corner of its body with each tentacle containing 500 000 stinging cells. That's 24 million stinging cells per animal!

Its toxin acts on the heart and can cause cardiovascular collapse and death as quickly as within 2-5 minutes.

Deaths occur each year all around the tropics; no data on the total number of fatalities has been recorded.

It all depends depend on how much venom is injected. A study from 1991-2004 found that of the 225 analysed cases of stings in Australia, 8% needed hospital treatment, 5% received anti-venom and there was a single fatality.

26% experienced severe pain, while it was moderate to none in the remaining 74%.

CARNIVOROUS CONE SHELLS

If you are a slow-moving animal and you feed on fish (relatively fast-moving animals), you need a fast-acting venom—a venom that will paralyse your prey in under a second to stop it swimming away. Welcome to the terrifying world of the cone shell.

Over 600 species occur, mostly in the tropics, and all are venomous. Most feed on worms and are very small animals, so thankfully a sting from one of these is like a mild bee sting— no harm there. However, the large, fish-eating species are potentially fatal to humans.

They have a developed tooth which they use as a harpoon; it is hollow. Attached to the base is a long tube which passes the toxin into the prey, and acts as a retrieval rope to pull in the paralysed victim.

Such is the toxic nature of the venom that when a fish gets hit, it is paralysed in a second and then pulled back to the snail for ingestion.

***Conus geographus*, the largest and deadliest cone shell, is also known as the 'cigarette snail' based on an exaggeration stating that when stung by this shellfish, the victim will have only enough time to smoke a cigarette before dying.**

Their brightly coloured shells are attractive to humans— most stings occur when picking them up. The harpoon is so powerful that a wetsuit or gloves are no protection from the deadly venom.

Two species are known to produce a type of insulin to cause hypoglycaemic shock in harpooned fish, paralysing them. They are the only two species known to use insulin as a natural weapon.

Ziconotide, a pain reliever recorded as 1000 times more powerful than morphine, was first isolated from the venom of a cone shell.

BLUE-RINGED MURDER

If you extract all the venom from a blue-ringed octopus and share it equally between 26 people, they will all be dead in 2 minutes.

There are 4 species of this very small octopus that measures no more than 20 cm. Thankfully, it remains out of the way of humans most of the time, being nocturnal. Its dull brown coloration soon transforms when exciting iridescent blue rings appear all over its arms. This warning signal was often ignored by humans as they wanted to explore the beast further— stings occurred.

Found mainly in the tropical regions of Australia (where else?) its distribution extends up to the Philippines and Japan, where the most northerly fatality happened.

The luckiest lady alive was filmed in 2019 with a blue-ringed octopus on her hand; it did not bite. The video went viral on TikTok.

No blue-ringed octopus anti-venom is available. The bite is painless, and you do not realise it until the venom's effects are felt. You must then wait and hope.

The main toxin is tetrodotoxin— the same as found in pufferfish. It is present in every cell of the octopus.

When laying eggs, the mother octopus will inject the toxin into her developing eggs to allow the young to synthesize their own venom on hatching.

A paralysed victim was rescued from death by emergency medical teams performing cardiopulmonary resuscitation on the beach. Paralysed, the poor casualty could not move their mouth to say that their eyelids were paralysed and stared at the sun until they were blinded.

Although the venom is very deadly, only 17 deaths have been recorded in Australia.

FLORAL AND LETHAL

"Instantly, I felt a severe pain resembling that caused by the cnidoblast of Coelenterata, and I felt as if the toxin were beginning to move rapidly to the blood vessel from the stung area towards my heart. After a while, I experienced a faint giddiness, difficulty of respiration, paralysis of the lips, tongue and eyelids, relaxation of muscles in the limbs, was hardly able to speak or control my facial expression and felt almost as if I were going to die. About 15 minutes afterwards, I felt that pains gradually diminish and after about an hour they disappeared completely. But the facial paralysis like that caused by cocainization continued for about six hours."

Tsutomu Fujiwara (1935)

The description above is the recorded effect of being stung by the flower urchin, the most dangerous member of the urchin family. This relatively large urchin of up to 20 cm is so-named as it appears to be covered in little flowers. Each flower packs a potent sting.

One unconfirmed death occurred when a pearl diver touched one; unable to reach the surface, the diver drowned.

FANTASTIC URCHIN FACTS

Clownfish will dance around the flower urchin, like an anemone, lightly brushing themselves against its

spines. This ensures they are exposed to a small amount of venom and are thus able to mimic the urchin and avoid more stings.

A new marine species was discovered on eBay! In 2006 British marine biologist, Dr. Simon Coppard, saw an unusual urchin for sale on eBay. From the photo, he was able to confirm that the species was unidentified, making it a new sea urchin discovery.

In the Pacific, bunches of urchin spines were used as sandpaper for shipbuilding.

TINY BUT TOXIC

Let us go down in scale and observe miniature beasts that produce such powerful toxins they can cause us a wide range of symptoms. The list includes blindness, memory loss, uncontrolled movement, paralysis, temperature reversal, vomiting, diarrhoea, plus many, many more vile symptoms—oh yes, and of course, death.

The culprits are bacteria living inside algal cells, which secrete various very powerful toxins with a wide range of effects. The ciguatera toxin can lodge itself in the body and last for over seven years, its effects only occurring when the person eats meat. So, for seven years every time you eat meat you are as sick as a dog— time to become a vegetarian, if you ask me.

It is estimated up to 1 million cases occur every year from eating contaminated shellfish; the vast majority result in time spent on the toilet and are not serious at all. Up to 50 000 cases of ciguatera occur each year from eating contaminated fish.

So how exactly do we succumb to toxins produced by such microscopic beasts? Well, it has much to do with the food chain and something known as bioaccumulation.

Let's look at shellfish poisoning. Bacteria live in phytoplankton and, as previously discussed, the phytoplankton bloom and suddenly there are huge amounts of food in the water—food for blue mussels and other filter feeders. Happy days for the shellfish, yet each algal cell they eat contains a tiny amount of bacterial toxin, which gets stored in their flesh. One mussel can ingest 50 million algal cells a day— that's a lot of toxins.

Along comes a cook and serves their prize shellfish dish. It's a dinner they have made hundreds of times, but not from shellfish exposed to a phytoplankton bloom. The tasteless toxin is consumed and soon its effects are realised.

Just what happens depends on the species of phytoplankton that bloomed; different species— different toxins—different symptoms.

Paralytic shellfish poisoning can be fatal in extreme cases. Symptoms can appear within ten minutes after consumption and include nausea, vomiting, diarrhoea,

141

abdominal pain, burning lips, face, neck, arms or legs.

Diarrhetic shellfish poisoning— the name indicates its symptoms: intense diarrhoea and abdominal pains along with being violently sick.

Amnesic shellfish poisoning causes death, brain damage or, if you're lucky, only permanent short term memory loss.

Neurotoxic shellfish poisoning again causes a wide range of symptoms, as described above with all the others. No fatalities have been recorded, but one symptom is quite unique— the sensation of ants crawling all over your body.

How to avoid such cases? Only purchase shellfish which have been purged. This means they have been placed in clean water for 14 days to remove the toxins, or they have been certified harvested from waters free of algal blooms.

Ciguatera can be fatal and is caused in the same manner, by eating fish that have bioaccumulated toxins from further down the food chain. Small fish eat algae, which is full of a bacterium secreting a toxin. The small fish is eaten, then that fish is eaten, and soon the big predator has stored up loads of the toxin. Along comes the fisherman, and in no time this fish is in the shop or restaurant and being eaten by some unsuspecting customer or tourist—pain will follow…

Symptoms may include diarrhoea, vomiting, numbness, sensitivity to hot and cold, dizziness, and weakness; the symptoms will go away only to rear their ugly heads again many months or even years later.

How do we avoid such illness? Well, the locals know which fish not to catch, and in areas where this is common, large fish are not eaten. But if you are unaware, it could be a lottery, where up to 50 000 people each year don't win!

THREATS TO THE OCEANS

The solution to pollution is dilution—an old saying in industry; however, we are now in a situation where the ocean is FULL UP.

This is the hardest thing I have ever had to write because it will detail the ongoing destruction that is affecting the life and habitats within our oceans. There is no fun here.

There are numerous television programs and books that examine this topic in great depth. Usually each subject is reported separately because each subject is HUGE and demands detailed coverage. Think of each theme as a sealed box with the following labels: plastics, reef destruction, overfishing, fishing methods, oil spillage, sewage disposal, waste disposal and habitat destruction.

Yet in reality, this is not what is happening. We must open all the boxes at once and let all their contents mix. Then we will expose the truth about the destruction that is going on in our waters on a global basis, day by bloody, polluting day.

THE FACTS OF DESTRUCTION

More oil reaches the oceans each year as a result of leaking automobiles and other sources than any major oil spill.

Air pollution is responsible for over 30% of the toxic contaminants that end up in oceans and coastal waters.

Longline fishing involves lines, often over 80 km (50 mi) long, with thousands of baited hooks. This activity is estimated to kill over 300 000 seabirds, including 100 000 albatrosses, on an annual basis.

Every day, over three times as much rubbish is dumped into the world's oceans as the weight of fish caught.

Less than 0.5% of marine habitats are protected -- compared with 12.5 % of global land area.

Every ecosystem has what is known as its carrying capacity. This is the natural number of animals and plants that an ecosystem can sustain in natural balance. The predator population will slowly increase over time, and with it, the prey population will decrease as more predators eat more prey. This goes on until it reaches a point where the predators start to die out as they cannot find enough food, and their population decreases. When this happens, the prey population increases, as less predators means more prey animals can reproduce. Once again, the predator population increases— you get the picture.

Everything is in balance, and the ecosystem progresses in a natural fashion. Unfortunately, humans do not fit this rule; our population has increased so much that the demands of our species are killing our planet— its carrying capacity has been exceeded.

It is estimated that over 75% of all pollution in seas and oceans comes from land-based activities.

Tropical coral reefs border the shores of 109 countries, the majority of which are among the world's least developed. Significant damage has been recorded in 93 countries.

The major causes of coral reef decline are coastal development, sedimentation, destructive fishing practices, pollution, tourism and global warming.

The least protected parts of the world are known as the high seas. These are areas beyond any government's control— they cover almost 50% of the earth's surface.

Although there are some treaties and fisheries agreements that protect ocean-going species such as whales, there are no protected areas in the high seas; they are regions of wanton destruction.

So, when did we as a species start to destroy the sea? The simple answer is at the beginning of the industrial revolution. Around the year 1750, carbon emissions started to rise, the human population started to grow and spread around the globe, while industrial and human waste was on the increase. However, it was not until the 1950's that we saw a huge increase in pollution and environmental destruction.

Only in the early 1960's did we began to realise and become concerned about the effect our activities were having on the planet and its wildlife as a whole.

COMPLICATED CHEMICALS

A chemical Dichlorodiphenyltrichloroethane (DDT) was widely used as a pesticide to protect crops from insect pests and to kill mosquitoes, preventing the spread of malaria. The solution to pollution is dilution did not work there. The insects were killed, and their bodies washed into rivers, or the spray entered rivers directly and was absorbed by invertebrates. These bugs were eaten by the fish or animals higher in the food chain and the DDT was stored in their bodies. If one fish eats one hundred invertebrates, it absorbs 100 molecules of DDT—an alarming picture. A massive number of fish deaths were reported and bird populations crashed as their eggs failed to hatch.

Enter Dr Rachel Carson (1907 -1964), a marine biologist who published the book Silent Spring (1962), detailing what could be happening to the wildlife of the land and sea as a result of that single chemical. Chemical companies issued lawsuits to stop it, but to no avail; through the publication of one book and her other writings, Dr Carson was singly responsible for advancing the environmental movement on a global basis. DDT was banned for many applications, although not entirely.

Her work cannot be underestimated; she was a true heroine of the natural world.

THE COST OF POLLUTION

An average of 600 000 barrels of oil per year has been accidentally spilled from ships.

Oil is a major polluter. However, it ignites a discussion between environmentalism and pure, hard science around the question: when is an event a disaster and when is it a disturbance? After an oil spill, we have all seen and been moved by the plight of the animals it smothers and kills.

One such event took place in the 1970's and a court ruled that a barnacle was worth 10 cents in compensation. Scientists went to the shore to work out the cost, only to be met by a healthy population of animals. With this in mind, the guilty party tried to avoid paying compensation, stating that everything was back to normal; it failed.

However, it did raise the argument that the spill was not a disaster but a biological disturbance, because in time the animals repopulated. Science has no feelings. I know which side of the fence I sit on and wonder if the defence lawyers would try this today!

OUT OF STOCK

Populations of large fish, such as tuna, cod, swordfish and marlin have declined by as much as 90%.

Accidently catching animals is called by-catch and is responsible for the mortality of small whales, dolphins, and porpoises— estimated to be more than 350 000 each year.

About 38.5 million tonnes of bycatch results from current preferred fishing practices each year. That's 38.5 million tons of dead animals thrown back into the sea!

At least 100 million sharks (figures show this could be too low) are killed each year for their meat and fins. The sharks are commonly de-finned while alive and thrown back into the ocean to die.

Fish consumption has risen by 50% since 1950, but the fish do not reproduce any faster!

More than 3.5 billion people depend on the ocean for their primary source of food. In 20 years, this number could double to 7 billion.

Norway, Japan and Iceland practice commercial whaling and kill over 100 000 small whales per year.

Conservation and fishing have been at loggerheads for decades, with the conservationists fighting against overfishing and harmful fishing practices and the fishermen fighting for what they can catch. This is an area where real balance needs to be addressed. On a global scale we have 3 problem areas, the developed countries, the underdeveloped countries and the destruction of 'legally caught marine mammals'.

There are responsible fishing nations where science gives the quotas, and fishermen can only take that amount. Also, fishing methods are improving by aiming to reduce by-catch, only fishing by line and changing net sizes to allow small fish to escape. Anyone can do their part by insisting on sustainably caught fish. Some fish sold in London and many other cities and towns will have the name of the boat and skipper recorded on each fish to ensure customers that it is from sustainable stock and line-caught to stop by-catch; surely this is the way forward?

Then we have the underdeveloped countries where overfishing and harmful fishing methods are major problems. We should not be too quick to condemn these people. Often, the majority are trying to feed their families and generate some sort of low income. They have nothing else in their lives; there is no other work—it is the only thing they can do. What is needed are conservation programmes that divert fishing efforts into other forms of income from the sea, such as aquaculture, of which there are many forms.

Overfishing is not the result of the majority of fishermen, who are as concerned as we all are. It is the result of greedy companies and individuals whose only motive is profit; sadly there are many.

Illegal and unregulated fishing constitutes an estimated 11-26 million tonnes (12-28%) of fishing world-wide.

Over 60% of fish stocks are fully-fished.

The world's largest fish producer and exporter is China, taking one third of the global production, while the EU is the world's largest importer of fish and fish products.

It is easy to repair fish stocks, and it is amazing how quickly they can recover— often in a matter of only a few years. 0.5% of the oceans are protected; this area needs to be enlarged, as marine reserves and the establishment of no-take zones have proved to be the fastest and most effective ways of allowing the natural balance to redress itself. If areas were set aside, fish stocks and all marine life would begin to repair and flourish.

RADIOACTIVE OCEANS

From 1946 to 1993, 13 countries dumped nuclear waste into the oceans. In 1993 this practice was banned.

To understand nuclear waste entirely, you need a degree in nuclear science; so I will attempt (as I do not have one!) to present the facts in a condensed manner.

There are 3 levels of waste: high, intermediate and low-level waste. High-level is the really hazardous one and consists of nuclear fuel waste. Intermediate waste is much, much less dangerous but still very bad, consisting of things like parts of reactors that have been exposed to high-level waste. Low-level waste comes from the medical, academic or industrial sectors.

Nuclear decay is measured in what is known as the half-life—the time taken for half the radioactive material to decay. So if a piece of uranium has 100 atoms, the time it takes these to decay into 50 atoms is its half-life.

Roughly (and I mean roughly), high-level waste has a half-life of over 24 000 years, so after this time half has decayed, but what is left has another 24 000 years remaining. Intermediate-level waste has a half-life of around 30 years and low-level has 5 years.

However, it is not just the half-life that matters, but also how much radiation is given off in one go. If you get hit by a tennis ball—that's it; however, if you get hit by 100 000 balls at once—that's a different matter. 10 tennis balls are equivalent to intermediate-level and the latter high-level.

The total amount of nuclear waste in the ocean will never be known because when it was dumped, it was poorly classified or not reported.

Only the Soviet Union dumped high-level waste (in the Artic); all the others dumped intermediate-level waste.

The Atlantic Ocean is located between the USA and Europe, so this is where the highest amount of waste is found.

The measurement for radioactivity is the becquerel (Bq), and waste is measured in trillions of becquerels. Don't be to be alarmed, as 1 becquerel is the decay of 1 atom, so very high numbers are the norm. A tetrabecquerel is a trillion becquerels, abbreviated as TBq.

The Soviet Union was the worst dumper, not only guilty of disposing of high-level waste, but of dumping around 39 000 TBq of waste. Next came the UK in a close second with around 35 000 TBq, and in bronze medal position was Switzerland, with 4 500 TBq. Surprisingly, the USA came fourth and, considering its size, dumped only 3 500 TBq.

Background radiation is the naturally-occurring level of radiation. The sea has quite a high level of background radiation, so it is known that leakage of nuclear waste would not raise this level, as the ocean is too vast.

However, the Artic sites containing high-level waste will be a different matter. If that waste is exposed, then all life around the area will cease.

TOXIC WATERS

The Citarum River in Indonesia is regarded as the most polluted in the world, receiving an estimated 20 000 tons of waste and 340 000 tons of waste water on a daily basis.

'All drains lead to the ocean' is a quote from the film Finding Nemo. Well, a more accurate quote would be 'most fresh water entering the sea is not fresh but polluted'.

Sitting here, I say to myself, 'Wow, where do I begin with this one?' Looking at a river, it starts in the mountains as a freshwater stream then meanders down through villages, getting bigger— the first waste is dumped. Then it flows through towns and the waste increases in cities and major industrial areas. Eventually, the river becomes a flow of contaminated water, laden with dissolved toxins and full of solid pollutants, garbage and human waste. The further it travels, the higher the degree of pollution, until it spills out into the ocean; 24 hours a day, every day. It's like someone being physically sick into a swimming pool, only it happens every second. Soon that swimming pool is not very attractive at all.

Around every 10 seconds, a child under the age of five dies from diseases related to contaminated water.

It is estimated that 2.5 million tonnes of sewage, industrial and agricultural waste are discharged into the world's oceans on a daily basis.

In developing countries, 70% of industrial waste is dumped untreated into waters. However, so-called developed countries like the USA and members of the European Community are just as guilty because the cost of cleaning waste is avoided to increase company profits.

Around 38% of lakes in America are too polluted for fishing.

It is estimated that 75% of all waste dumped into rivers comes from untreated sewage, and 80% of river discharge into the Mediterranean Sea is sewage.

Water discharges into the oceans mainly through rivers, but any coastal areas may also have pipelines reaching into the sea. The major pollutants can be classified into 3 broad areas: organic, chemical, and solid waste.

Organic waste is mainly sewage (often untreated) and any other waste that will biodegrade, such as paper and the bodies of animals. These waste products release deadly bacteria, which rots the waste, using up oxygen supplies and making the water very toxic. The waste sinks and smothers the bottom, so that only a few species can live there. This vastly reduces the biodiversity of not just the sediment-loving animals, but also the predators that feed upon them.

Farming and industrial waste contain countless chemicals that enter the waters via different methods. Many of these chemicals are deadly and accumulate to high levels over time, often concentrating along the food chain like the previously-mentioned DDT. The whole area is depleted of animals as they die out.

Solid waste is often garbage— anything you would find in your domestic bin—just discarded into the water. All of this becomes concentrated into higher and higher levels as the river flows further and is finally discharged into the sea.

DEAD ZONES

If you drop a ton of polluting chemicals into water, the effects are reduced the further you move away by the dilution effect. If you drop a ton of pollution into the same spot every second then the centre spot becomes so concentrated that all life ceases to exist there; that deadly centre spot is now increasing in size every day. Here we enter the realm of dead zones.

Dead zones are areas of water, both fresh and salty, that cannot support life, due to pollution. It is estimated that there are over 500 dead zones globally. If you add them up you will have a dead zone the size of the UK.

The largest single dead zone is found in the Gulf of Mexico, caused by the Amazon River and its very high nitrogen levels. It grows and reduces in size every year depending on the amount of waste discharged. At its peak, it covered 13 500 km^2 (8 500 sq.mi).

As with overfishing, cleaning up rivers and their effects on the oceans is easy. Yes, it costs money, but surely the companies that produce the waste should treat it? I wish!

Rivers receive fresh water daily; this water would flush out all the pollutants quite quickly if they were prevented from being added. The River Mersey in northwest England was once a dirty dead river, due to its passage through many industrial towns. Since 1980 this river received less and less pollutants, thanks to legislation and public concern. Within 20 years you could see anglers on the banks of the Mersey catching lots of fish, including large ones, and salmon even run in the river again. This is fantastic for riverine wildlife and also for the Irish Sea, where it discharges fresh water rather than slime.

PERVASIVE PLASTIC

Close your eyes and think of holidaying in the Maldives, a tropical paradise by anyone's standard. Then think of the single island that is now the waste disposal dump— all the holidaymakers' waste has to

go somewhere. Sewage flows into the reefs and solids are dumped on an island. Not such a paradise now!

90% of plastic waste dumped into the oceans comes from just 10 rivers.

Plastic waste kills up to 1 million sea birds, 100 000 sea mammals and countless fish each year.

The Great Pacific Garbage Patch is caused by converging surface currents from the USA and Asia, which carry surface-floating plastic waste. It contains 2 trillion pieces of plastic weighing 80 000 tons. When you realise that 70% of plastics sink to the bottom of the sea, this fact gets even more frightening. Out of sight, out of mind!

When we talk about plastic waste, I often think that the dangers to marine life are not truly conveyed. Its potentially devastating effects are quite profound. Think of a piece of plastic netting, just a small piece floating around. A fish is attracted to it and soon becomes entangled, swimming erratically as if injured. A predator picks up the vibrations and homes in on the fated fish. Soon the fish is ingested along with the plastic, causing the predator to die. Two lives lost— one piece of plastic!

Now the dead predator sinks, and its body is broken up by scavengers such as crabs. Part of the plastic is taken into their bodies as well. 10 crabs die. So 12 lives are lost due to one piece of plastic!

Once the predator's body has been completely eaten and all that is left is the original plastic netting that caught the first fish, it won't decay. Currents move the netting up into the water column and soon a fish is caught, a predator eats it and so on— you get the picture. One piece of plastic can kill well over 100 animals again and again and again.

STARK PLASTIC FACTS

96% of disposable nappies that have ever been made are still in existence today; others have been burnt to pollute the atmosphere.

Depending on the type of plastic, it takes an average of 500 to 1000 years to degrade.

Over 100 million marine animals die each year from plastic pollution.

If you weighed the total human population on earth— that's the weight of one years' worth of plastic production.

85% of sea birds have been contaminated with plastic.

Plastic is here to stay full stop. It's how we manage it that will make the difference. Governments, companies and individuals have an effect. In 2015 the UK imposed a 5p charge for a single use carrier bag in retail shops. This was to encourage the use of bags that are retained and used many times, thus reducing the amount sent to landfill and, ultimately, the oceans. From 2017-2018 there were 1.75 billion bags issued, in 2018-2019 this had dropped to 1.11 billion— it works.

Ghost fishing happens when lost fishing nets, drifting in the sea, entangle marine mammals such as whales and dolphins, causing them to drown. Over 700 000 tonnes of these nets are estimated to be floating and killing in the oceans today.

Cigarette filters are the most common item collected from beach pickups. They have a 30-year life span and are eaten by fish, blocking their insides.

China tops the list as the most polluting nation, with nearly 9 million tonnes of mismanaged plastics, followed by Indonesia with over 3 million; there's a huge gap between first and second place.

So what drives this plastic pollution? Money. Cheap production costs are passed onto the consumer, so we are all guilty.

Fortunately, as with climate change, awareness is growing and people are changing their behaviour. Wooden toothbrushes, paper straws and recyclable plastics are now more common, along with biodegradable plastics. What is currently in the oceans will remain there until it is removed (but who will pay for it?). Beach clean-ups are now more common, and littering is looked down upon.

Yes, we are going in the right direction, but as with climate change reversal, it will take a long time and sadly, more animals are doomed.

LIGHT AND SOUND

There are two types of pollution that have been going on for many years, but only in the last 50 years have they really started to distress animals and cause fatalities. They are not widely reported as there is basically very little scientific data; but we know it happens— light and sound pollution.

A study by Plymouth University found light in the blue and green spectrum from LED road lights reached the ocean floor far out to sea. This disrupted the life cycles of animals relying on light to guide their daily activities.

Cities and towns located near beaches where turtles hatch can cause turtle babies to crawl towards roads, guided by the lights, which they mistake for the moon that would lead them to the sea.

Noise and more noise— well, that's the ocean; it is a very loud place indeed. It's not just the volume, but the type of sound, which travels thousands of kilometres. Human activities have disrupted the natural balance beyond belief.

A study has identified that sound can disrupt the ability of larvae to find their natural adult habitats, resulting in huge losses to the adult population.

A high number of animals (not just marine mammals) use sound to meet and greet, communicate, and live a natural life; this is being disturbed on a global scale.

During COVID lockdown, marine noise was reduced by up to 20%. The near-immediate result was observed when mammals appeared in locations where they had never been seen before. Whale calves were observed at far greater distances from their mothers, as the parents could hear them from further away.

To reverse the effects of pollution will take many, many years but we will see the effect of noise reduction in months!

THE CLIMATE CONUNDRUM

Global warming is driving climate change, which is potentially the greatest destructive force since the meteor hit Mexico 65 million years ago and caused a mass extinction of life on earth.

Three of the human race's biggest allies in fighting climate change are phytoplankton, mangrove forests and the seagrass meadows. They convert 10 times more atmospheric carbon dioxide than all land-based plants.

Yet, with the same zeal as they are deforesting tropical rainforests, humans are destroying mangroves and seagrass at an alarming rate.

Which idiots would attack their main allies in a war? The war against climate change!

Only humans. Full stop.

There is a well-known phrase related to the polar seas, 'it's the tip of the iceberg,' and that's exactly what this book has been about. The amazing facts that we have discussed and the detailed explanations are indeed just the tip of the iceberg. There is so much more out there and so much we have yet to discover. I only hope you have had as much enjoyment reading this book as I did writing it; and to those who go on to study further, those who clean beaches, those who help in any way whatsoever, the oceans thank you, because I think we all agree—Oceans are indeed incredible.

The Marine Life Series

A Four-book series by Andrew Caine

Marine Biology for the Non-Biologist

THE BASIS OF LIFE IN THE OCEANS

SOFT-BODIED ANIMALS - THE CNIDARIANS
THE JELLYFISH
THE HYDROIDS
THE ANEMONES
THE CORALS
THE STONY CORALS

SHELLFISH - THE MOLLUSCS
THE GASTROPODS
THE MESOGASTROPODS
THE ARCHEOGASTROPODS
THE NEOGASTROPODS

THE NUDIBRANCHS (SEA SLUGS)
THE BIVALVES
THE CEPHALOPODS
THE NAUTILUS
THE SQUIDS, CUTTLEFISH AND OCTOPI

ANIMALS WITH EXOSKELETONS - THE CRUSTACEANS
THE DECAPODS- CRABS, LOBSTERS AND SHRIMPS
THE BARNACLES

ANIMALS WITH SPINY SKINS - THE ECHINODERMS
THE STARFISH
THE BRITTLE STARS
THE SEA URCHINS
THE SEA CUCUMBERS
CORAL REEF ARCHITECTURE

MARINE INVERTEBRATE TOXINS
LIMU-MAKE-O-HANA (THE DEADLY SEAWEED OF HANA)
SHELLFISH POISONING
PARALYTIC SHELLFISH POISONING (PSP)
NEUROTOXIC SHELLFISH POISONING (NSP)
DIARRHOEIC SHELLFISH POISONING (DSP)
AMNESIC SHELLFISH POISONING (ASP)
CIGUATERA
THE CONE SHELLS
SEA SNAKES
VENOMOUS FISH

HYDROTHERMAL VENTS AND VENT BIOLOGY

THE DISCOVERY
THE PHYSICAL ENVIRONMENT
VENT BIOLOGY
LIFE IN THE POLAR SEAS

THE POLAR ENVIRONMENT
ANIMAL ADAPTATIONS
THE FISH
THE INVERTEBRATES
THE BIRDS
THE MAMMALS

Marine Ecology for the Non-Ecologist

ECOLOGY – THE BASIC FACTS

THE PHYSICAL ENVIRONMENT
THE BIOLOGICAL ENVIRONMENT
FOOD CHAINS, WEBS AND ENERGY FLOW
RANDOM 1 – THE IMMORTAL JELLYFISH

THE PHYSICAL ASPECTS THAT SHAPE THE COASTAL ENVIRONMENT

WATER MOVEMENT
WAVES, TIDES AND CURRENTS
TEMPERATURE
SALINITY

RANDOM 2 – THE BLUE DRAGON

THE ROCKY SHORE

THE PHYSICAL ENVIRONMENT
HIGH AND LOW ENERGY AREAS
ZONATION
THE ALGAE
THE ANIMALS
BEHAVIOUR
MINI HABITATS
RANDOM 3 – SAMANTHA THE SERPENT STAR

THE MUD AND THE SAND

THE SHAPE OF THE SHORE
GRAIN SIZE
THE SEDIMENTS
SEAGRASS
THE ANIMALS
RANDOM 4 – THE EMPTY SEA

WHERE RIVERS MEET THE SEA

TYPES OF ESTUARIES
THE SALT WATER: FRESH WATER MIX
SHAPES OF ESTUARY
ANIMALS, ALGAE AND PLANTS
RANDOM 5 – SAR 11 THE MOST ABUNDANT LIFE FORM IN THE OCEANS

THE MARSHES AND THE MANGROVES

ESTABLISHMENT
ZONATION
PRODUCTIVITY
ANIMALS, ALGAE AND PLANTS

RANDOM 6 – BERNADETTE THE DEADLY YELLOW BOXFISH

CORAL REEFS

DISTRIBUTION AND LIMITING FACTORS
THE THREE REEFS
FORMATION OF THE CORAL REEF
ZONATION
ANIMALS AND ALGAE
RANDOM 7 – SID THE SPONGE

INDEX

RANDOM 8 – SIR ISAAC NEWTON AND FLUORESCENT CORALS

A Marine Biology Students Journal

My Marine Biology Journey

The first and only journal for the marine biology student or anyone who is interested in this wonderful world of science, that is full of essential facts each student must know.

Not just a book of empty lined pages.

This journal has over 100 facts that every marine biology student should understand one on every page.

A5 size makes it an ideal size to pack into your travel bag but also fantastic when sitting down and reflecting on the day.

Examples of headline facts are:-

Polychaeta, the bristle worms or polychaetes, are a paraphyletic class of annelid worms. Each body segment has a pair of fleshy protrusions called parapodia that bear many bristles, called chaetae.

Cnidarian's such as sea anemones and corals form a symbiotic relationship with a class of dinoflagellate algae, zooxanthellae of the genus Symbiodinium.

Halophyte is a word to describe plants that are salt tolerant such as mangrove trees.

The internal organs of an urchin are enclosed in a hard shell called the test, it is composed of fused plates of calcium carbonate covered by a thin dermis and epidermis.

Printed in Great Britain
by Amazon